Praise for *Who's in Charge?*

"A fascinating, accessible, and often humorous read for anyone with a brain! And a must read for neuroscientists, psychologists, psychiatrists, and criminal attorneys."   —*Library Journal* (starred review)

"Gazzaniga is a giant in cognitive neuroscience. . . . [He] advances a fascinating argument that both limits and contextualizes brain research."   —Forbes.com

"A fascinating affirmation of our essential humanity."

—*Kirkus Reviews*

"Gazzaniga is calling for a new 'vocabulary'—one that doesn't yet exist—that captures the dynamism between the brain and the mind. . . . And no one is better positioned than Gazzaniga to speak to this point. . . . He challenges the concept of free will as we know it."

—Daily Beast

"An utterly captivating and fascinating read that addresses issues of consciousness and free will and, in the end, offers suggestions as to how these ideas may or may not inform legal matters."

—*Daily Texan*

"Gazzaniga seems to have lost none of his youthful excitement. . . . One of the many rewards of *Who's in Charge?* is a compelling account of the evolution of our hypermodularity."   —*Nation*

"Deliver[s] an up-to-date review of the research in psychology and neuroscience. . . . Take[s] us on a journey into the machinery of human decision-making, its constrictions and flaws. By studying how we make choices, we can learn to make better ones."   —*Nature*

MICHAEL S. GAZZANIGA

ecco

An Imprint of HarperCollins*Publishers*

WHO'S IN CHARGE?

FREE WILL AND
THE SCIENCE OF THE BRAIN

Illustration on page 79 taken from the book *Mind Sights* by Roger N. Shepard. Reprinted by permission of Henry Holt and Company, LLC.

HarperCollins books may be purchased for educational, business, or sales promotional use. For information please e-mail the Special Markets Department atSPsales@harpercollins.com.

A hardcover edition of this book was published in 2011 by Ecco, an imprint of HarperCollins Publishers.

FIRST ECCO PAPERBACK EDITION PUBLISHED 2012.

The Library of Congress has cataloged the hardcover edition as follows:

Who's in charge? : free will and the science of the brain / Michael S. Gazzaniga.—1st ed.
    p. cm
  ISBN 978-0-06-190610-7 (hardcover)
  ISBN 978-0-06-209683-8 (e-book)
  1. The way we are   2. The parallel and distributed brain
  3. The interpreter   4. Abandoning the concept of free will
  5. The social mind   6. We are the law   7. An afterword.
    Includes bibliographical references and index.

  QP360.5 .G396 2011
  612.8/233
                                        2011010015

ISBN 978-0-06-190611-4 (pbk.)

12 13 14 15 16 OV/RRD 10 9 8 7 6 5 4 3 2 1

For Charlotte,
Unquestionably the Eighth Wonder of the World

CONTENTS

INTRODUCTION

FROM SCOTLAND, FOR MORE THAN 125 YEARS, THE GIFFORD Lectures have been dispatched into the world owing to the behest and endowment of Adam Lord Gifford, a nineteenth-century Edinburgh advocate and judge with a passion for philosophy and natural theology. Under the terms of his will he directed these lectures to be given on the topic of natural theology with the stipulation that the subject be treated "as a strictly natural science" and "without reference to or reliance upon any supposed special exceptional or so-called miraculous revelation. I wish it considered just as astronomy or chemistry is. . . . [T]hey may freely discuss . . . all questions about man's conceptions of God or the Infinite, their origin, nature, and truth, whether he can have any such conceptions, whether God is under any or what limitations, and so on, as I am persuaded that nothing but good can result from free discussion." The lectures have focused on religion, science, and philosophy. If you have truly sampled the books that have flowed from them, you will quickly discover their bone-rattling quality. Some of the greatest minds of the Western world have delivered their ideas during the course of these lectures such as William James, Niels

Bohr, and Alfred North Whitehead to mention a few. Many of the long list of participants have precipitated major intellectual battles; some have spelled out the vastness of the universe or decried the failure of the secular world to provide a hopeful message about the meaning of life, while others have flat out rejected theology—natural or otherwise—as a worthwhile topic for grown-ups to spend any time thinking about. Seemingly, everything has been said, and all of it is stated with such clarity and force that when the assignment fell to me to add my own perspective, I almost withdrew.

I think I'm like everyone who has read the many books that have come from those lectures. We all feel the tug of an insatiable desire to carry on the quest to know more about the situation in which we humans find ourselves. In a way, we are stupefied by our interest because we do now know a great deal about the physical world, and most of us believe the implications of our modern knowledge, even though we sometimes have a hard time accepting wholly scientific views. Thinking about these things is what a Gifford lecture is all about, and I found myself wanting to throw in my own two cents' worth. Though submitting my own perspective in that forum is as scary as it is heady, I do want to show that all of the spectacular advances of science still leave us an unshakeable fact. *We are personally responsible agents and are to be held accountable for our actions, even though we live in a determined universe.*

We humans are big animals, clever and smart as we can be, and we frequently use our reasoning to a fault. And yet, we wonder, is that it? Are we just a fancier and more ingenious animal snorting around for our dinner? Sure, we are vastly more complicated than a bee. Although we both have automatic responses, we humans have cognition and beliefs of all kinds, and the possession of a belief trumps all the automatic biological process and hardware, honed by evolution, that got us to this place. Possession of a belief, though a false one, drove Othello to kill his beloved wife, and Sidney Carton to declare, as he voluntarily took his friend's place at the guillotine, that it was a far, far better thing he did than he had ever done. Humans are the last word, even though we can feel occasionally pretty inconsequential as we look up at the billions of stars and universes within which we are situated. The question still haunts us, "Are we not part of a bigger scheme of meaning?" Conventional hard-earned wisdom from science and much of philosophy would have it that life has no meaning other than what we bring to it. It's completely up to us, even though the gnawing question always follows as to whether or not that really is the way it is.

But now, some scientists and philosophers are even suggesting that what we bring to life is not up to us. Here are some truths of modern knowledge and their awkward implications. The physiochemical brain does enable the mind in some way we don't understand and in so doing, it follows the physical laws of the universe just like other matter. Actually, when we think about it, we wouldn't want it any other way. For instance, we wouldn't want our actions, such as lifting our hand to our mouth, to result in a random movement: We want that ice cream in our mouth not on our forehead. Yet, there are those who say that because our brains follow the laws of the physical world, we all, in essence, are zombies, with no volition. The common assumption among scientists is that we know who and what we are only after the fact of nervous system action. Most of us, however, are so busy, we can't take time out to think through or be burdened by such claims, and only a few of us succumb to existential despair. We want to do our jobs, get home to our wife or husband and kids, play poker, gossip, work, have a Scotch, laugh about things, and simply live. We seemingly don't puzzle the meaning of life most of the time. We want to live life, not think about it.

And yet, a certain belief is palpably dominant in the intellectual community, and that belief is that we live in a completely determined universe. This belief seems to logically follow from all that our species has learned about the nature of the universe. Physical laws govern the happenings in the physical world. We are part of that physical world. Therefore, there are physical laws that govern our behavior and even our conscious self. Determinism reigns—both physical and social—and we are asked to accept it, and to move on. Einstein bought it. Spinoza bought it. Who are we to question it? Beliefs have consequences and indeed, because we live in what is believed by many to be a determined world, we are commonly asked to be slow to assign blame and to not hold people accountable for their actions or anti-social behavior.

Over the years, Gifford Lecturers have approached the issue of determinism from many different perspectives. The quantum physicists have said there is wiggle room on the idea of determinism ever since quantum mechanics replaced the Newtonian view of matter. There is uncertainty at the atomic and molecular level, and this fact means you are free to choose the Boston cream pie over the berries the next time the dessert tray is passed around; your choice was not determined at the very instant of the big bang.

At the same time, others have argued that atomic uncertainties are not relevant to the workings of the nervous system and how it ultimately produces the human mind. The dominant idea in modern neuroscience is that a full understanding of the brain will reveal all one needs to know about how the brain enables mind, that it will prove to be enabled in an upwardly causal way, and that all is determined.

We humans seem to prefer black and white answers to questions, binary choices, all or nothing, all nature or all nurture, all determined or all random. I will argue that it is not that simple and that modern neuroscience is not, in fact, establishing what amounts to a wholesale fundamentalism with respect to determinism. I will maintain that the mind, which is somehow generated by the physical processes of the brain, constrains the brain. Just as political norms of governance emerge from the individuals who give rise to them and ultimately control them, the emergent mind constrains our brains. In a time when we all think we can agree that causal forces are the only way to understand our physical world, are we not in need of a new frame of thinking to describe the interactions and mutual dependence of the physical with the mental? As Professor John Doyle at Caltech points out, in the world of hardware/software, where everything is known about both systems, their functionality only exists by both realms interacting. Yet, no one has captured how to describe that reality. Something like the big bang happened when mind emerged from the brain. Just as traffic emerges from cars, traffic does ultimately constrain cars, so doesn't the mind constrain the brain that generated it?

Like trying to sink a cork in water, the issue won't go away. It keeps popping back. How the mind relates to brain, with its implications for personal responsibility, no matter who addresses it, keeps grabbing our attention. The importance of the answer to this question, which is central for understanding what we humans are experiencing as sentient, forward-looking, and meaning-seeking animals cannot be overstated. I wish to continue in the tradition that examines this fundamental issue and to outline the progress, as I see it, on how that interface of mind and brain might best be understood. Does the mind constrain the brain, or does the brain do everything from the bottom up? It's tricky, because in nothing that follows here am I suggesting the mind is completely independent from the brain. It is not.

In starting our journey it is important to review what sort of creatures we think we know we are in the twenty-first century. During the last one hundred years there has been a massive accumulation of knowledge about what makes us tick. It is truly daunting, and the question before us now is, Has it trumped earlier understandings on the nature of the human existence?

In my Gifford Lecture series and in this book, I see it as my duty to review the human knowledge of our time that many of the great minds of the past did not possess. Even with all of the fantastic comprehension gained about the mechanisms of mind that neuroscientists now have worked out, none of it impacts responsibility—one of the deep core values of human life. In substantiating this claim, I am going to explain the route and some of the detours that we have taken to reach our current knowledge of the brain and review what we currently know about how it works. To understand some of the claims that have been made about living in a deterministic world, we will visit a few different layers of science, going from the micro world of subatomic particles, places you never thought neuroscience would take you, to the macro social world of you and your buddy high-fiving over the Super Bowl game. These wanderings are going to show us that the physical world has different sets of laws depending on what organizational layer one is looking at, and we will discover what that has to do with human behavior. We are going to end up, of all places, in the courtroom.

Even with all the knowledge of physics, chemistry, biology, psychology, and all the rest, when the moving parts are viewed as a dynamic system, there is an undeniable reality. We are responsible agents. As my kids say, "Get over it." Human life is a really good deal.

Chapter One

THE WAY WE ARE

THERE IS THIS PUZZLE ABOUT EVERYDAY LIFE: WE ALL FEEL like unified conscious agents acting with self-purpose, and we are free to make choices of almost any kind. At the same time everyone realizes we are machines, albeit biological machines, and that the physical laws of the universe apply to both kinds of machines, artificial and human. Are both kinds of machines as completely determined as Einstein, who did not believe in free will, said, or are we free to choose as we wish?

Richard Dawkins represents the enlightened science view that we are all determined mechanistic machines and immediately points out an implication. Why do we punish people who engage in antisocial behavior? Why don't we simply view them as people who need to be fixed? After all, he argues, if our car stalls and fails us, we don't beat it up and kick it. We fix it.

Switch out the car for a horse that bucked you off. Now what do we do? The thought of a good poke does pop into the mind more than a trip to the barn for repairs. Something about animate flesh calls upon a seemingly vibrant set of responses that are part of us humans and pull along with them a host of feelings and values and goals and intentions and all those human mental states. In short, there is something about the way we are built, and presumably our brains, that appears to be governing a lot of our everyday behavior and cognition. We seem to have a lot of complexity in our makeup. Our very own brain machine runs on its own steam, even though we think we are in charge. Now that is a puzzle.

Our brains are a vastly parallel and distributed system, each with a gazillion decision-making points and centers of integration. The 24/7 brain never stops managing our thoughts, desires, and bodies. The millions of networks are a sea of forces, not single soldiers waiting for the commander to speak. It is also a determined system, not a freewheeling cowboy acting outside the physical, chemical forces that fill up our universe. And yet, these modern-day facts do not in the least convince us there is not a central "you," a "self" calling the shots in each of us. Again, that is the puzzle, and our task is to try and understand how it all might work.

The accomplishments of the human brain are one good reason we are convinced of our central and purposeful self. The modern technology and know-how of humans is so crazy-amazing that a monkey with a neural implant in North Carolina can be hooked up to the Internet, and, when stimulated, the firing of his neurons can control the movements of a robot in Japan. Not only that, the nerve impulse travels to Japan faster than it can travel to that monkey's own leg! Closer to home, take a look at your dinner. If you are lucky, tonight you might have a locally grown salad with sliced pears from Chile and an amazingly tasty gorgonzola from Italy, a lamb chop from New Zealand, roasted potatoes from Idaho, and red wine from France. How many different creative and innovative people cooperated in both scenarios to pull them off? Tons. From the person who first thought about growing his own food, and the one who thought the old grape juice was a bit interesting, to Leonardo, who first drew a flying machine, to the person who took the first bite of that moldy-looking cheese and thought they had a winner, to the many scientists, engineers, software designers, farmers, ranchers, vintners, transporters, retail dealers, and cooks who contributed. Nowhere in the animal kingdom does such creativity or cooperation between unrelated individuals exist. Perhaps even more amazing is that there are people who do not see much difference in the abilities of humans and that of other animals. In fact, they are pretty sure that their darling dog with the big, sad eyes is just a hair's breadth away from getting his self-help article published: "How to Manipulate Your Human Housemate Without Even Getting Off the Couch."

Humans have spread across the world and live in hugely varying environments. Meanwhile, our closest living relatives, the chimps, are endangered. You have to ask why humans have been so wildly successful, while our closest living relations are barely hanging on. We can solve problems that no other animal can solve. The only possible answer is that this came about because we have something that they do not. Yet we find this difficult to accept. As we are perched here at the beginning of the twenty-first century, we have more information to help answer some of these questions, information that was not available to the curious and inquiring minds of the past. And curious were those who have gone before us: Human interest in what and who we are is at least as old as history. Etched in the walls of the seventh-century B.C. Temple of Apollo in Delphi is the advice KNOW THYSELF. Man has always been intrigued with the nature of the mind, self, and the human condition. Where does this curiosity come from? That is not what your dog is thinking about on the couch.

Today, neuroscientists are exploring the brain by poking it, recording from it, stimulating it, analyzing it, and comparing it with those of other animals. Some of its mysteries have been revealed and theories abound. Before we get all impressed with our modern selves, we need to keep our egos in check. Hippocrates, in the fifth century B.C., wrote as if he were a modern neuroscientist: "Men ought to know that from nothing else but the brain come joys, delights, laughter and sports, and sorrows, griefs, despondency, and lamentations. And by this . . . we acquire wisdom and knowledge, and see and hear, and know what are foul and what are fair, what are bad and what are good, what are sweet and what unsavory. . . . And by the same organ we become mad and delirious, and fears and terrors assail us. . . . ."[1] His mechanisms of action were sketchy, but he had the principles down.

So I guess that leaves science to explain the mechanisms, and in doing so we best take the advice of Sherlock Holmes, who was known for his scientific method: "The difficulty is to detach the framework of fact—of absolute undeniable fact—from the embellishments of theorists and reporters. Then, having established ourselves upon this sound basis, it is our duty to see what inferences may be drawn and what are the special points upon which the whole mystery turns."[2]

This impulse, just nothing but the facts, is a way to start solving a puzzle, and early brain scientists started in that spirit. What is this thing? Let's get a corpse, open up the skull, and take a look. Let's make holes in it. Let's study people with stroke. Let's try to record electrical signals from it. Let's see how it hooks itself up during development. As you will see, those are the sort of simple questions that motivated early scientists and still motivate many today. As I go through our story, however, it will become evident that without actually studying the behavior of organisms or knowing what our evolved mental systems were selected to do, settling the question of "self" versus machine becomes a hopeless goal. As the great brain scientist David Marr observed, there is no way to understand how a wing of a bird works by studying its feathers. As the facts accumulate, we need to give them functional context and then examine how that context may, in fact, constrain the underlying elements that generate the function. Let's begin.

BRAIN DEVELOPMENT

Something short and snappy-sounding like "brain development" should be simple to study and understand, but in humans development ranges far; it takes in not only the neural, but also the molecular, and not only cognitive change over time, but also the influence of the external world. It turns out not to be simple at all: Oftentimes detaching the framework of fact from the theorizing is a long and arduous process with many detours, and such was the fate of unraveling the basics of how the brain develops and works.

EQUIPOTENTIALITY

The early twentieth century had suffered such a detour, the repercussions of which, both in the scientific and lay worlds, are still plaguing us in the form of the nature-versus-nurture question. In 1948, at my alma mater, Dartmouth College, two of Canada's and America's great psychologists, Karl Lashley and Donald Hebb, came together to discuss the following question: Is the brain a blank slate and largely what we call today "plastic," or does the brain come with constraints and is it somewhat determined by its structure?

At the time, the blank slate theory had reigned for the previous twenty years or so, and Lashley had been one of its early proponents. He was one of the first researchers to employ physiological and analytical methods to study brain mechanisms and intelligence in animals; he had carefully induced damage to the cerebral cortex in rats and quantified it, measuring their behavior before and after he made the lesions. While he found that the amount of cortical tissue he removed affected learning and memory, the location of it did not. This convinced him that the loss of skills was related to the volume of excised cortex rather than its location. He did not think that a specific lesion would result in the loss of a specific ability. He proposed the principles of *mass action* (the action of the brain as a whole determines its performance) and *equipotentiality* (any part of the brain can carry out a given task, thus no specialization).[3]

Lashley, while doing his graduate studies, had come under the influence, and became a good friend, of John Watson, the director of the psychological laboratory at Johns Hopkins University. Watson, an outspoken behaviorist and "blank-slater" famously said in 1930, "Give me a dozen healthy infants, well-formed, and my own specified world to bring them up in and I'll guarantee to take any one at random and train him to become any type of specialist I might select—doctor, lawyer, artist, merchant-chief and, yes, even beggar-man and thief, regardless of his talents, penchants, tendencies, abilities, vocations, and race of his ancestors."[4] Lashley's principles of mass action and equipotentiality fit well within the framework of behaviorism.

More evidence for this idea of equipotentiality came from one of the first developmental neurobiologists, Paul Weiss. He also thought that the brain was not that specific in its development and coined the famous phrase, "function precedes form,"[5] based on the results of his experiments in which he grafted an additional limb onto a newt, an amphibian in the salamander family. The question was, Did the nerves grow out to the limb specifically or did the nerves grow out randomly and then through the use of the limb become adapted to be limb neurons? He had found that transplanted salamander limbs would become innervated and capable of learning movement that was fully coordinated and synchronized with the adjacent limb. Roger Sperry, Weiss's student and later my mentor, summarized Weiss's widely accepted resonance principle as "a scheme in which the growth of synaptic connections was conceived to be completely nonselective, diffuse, and universal in downstream contacts."[6] So at the time it was thought that "anything went" in the nervous system—(neuron to neuron) there was no structured system. Lashley started it, the behaviorists pushed it, and the greatest zoologist of the time agreed.

NEURONAL CONNECTIONS AND NEUROSPECIFICITY

But Donald Hebb was not convinced. Although he had studied with Lashley, he was an independent thinker and started to develop his own model. He began to think that it was how specific neuronal connections worked that was important and shied away from the ideas of mass action and equipotentiality. He had already rejected the ideas of Ivan Pavlov, the great Russian physiologist, who had seen the brain as one big reflex arc. He was convinced that the operations of the brain explained behavior, and that psychology and biology of an organism could not be separated, a well-accepted idea now, but unusual at the time. Contrary to the behaviorists who thought that the brain merely reacted to stimuli, he recognized that the brain was always running, even when there was no stimulus present. He strove for a framework that captured that fact with the limited data on brain function that was available in the 1940s.

Hebb set about to postulate how this occurred based on his research. The death knell for strict behaviorism and the return to an earlier idea of neural connectivity's being of great importance came in 1949 with the publication of Hebb's book *The Organization of Behavior: A Neuropsychological Theory*. He wrote: "When an axon of cell *A* is near enough to excite a cell *B* and repeatedly or persistently takes part in firing it, some growth process or metabolic change takes place in one or both cells such that *A*'s efficiency, as one of the cells firing *B*, is increased."[7] Colloquially this is known in neuroscience as "Neurons that fire together, wire together" and forms the basis of Hebb's proposals for learning and memory. He proposed that groups of neurons that fire together make up what he called a *cell assembly*. Neurons in the assembly can continue to fire after an event that has triggered them, and he suggested that this persistence is a form of memory and that thinking is the sequential activation of assemblies. In short, Hebb's ideas pointed out the centrality of the idea of the importance of connectivity. It remains a central topic of study in neuroscience today.

Hebb focused his attention on neural networks and how they might work to learn information. While he did not focus on how those networks came to be, one of the implications of his theory is that thinking affects the development of the brain. In fact, in earlier experiments on rats published in 1947, Hebb had shown that experience can affect learning.[8] Hebb understood that his theory would undergo revision as more discoveries about brain mechanisms were made, but his insistence on combining biology with psychology marked the path that in little more than a decade led to the new field of neuroscience.

It was beginning to be understood that once information was learned and stored, specific brain areas had used that information in different, particular ways. The question remained, however, how did the networks form? In short, how does the brain develop?

The foundational work that became the backbone of modern neuroscience and emphasized the importance of neurospecificity was done by Paul Weiss's student Roger Sperry. How the connectivity, or wiring, took place was the question that fascinated him. He was skeptical of Weiss's explanation of nerve growth, where functional activity played a predominant role in the formation of neural circuitry. In 1938, the year that he began his research, other rumblings against the doctrine of the functional plasticity of the nervous system came from two Johns Hopkins Medical School physicians, Frank R. Ford and Barnes Woodall, when they recounted their experiences with clinical patients whose disorders of function, after nerve regeneration, persisted for years without improvement.[9] Sperry set out to investigate functional plasticity in rats by seeing what the behavior effects were of changing nerve connections. He switched the nerve connections between the opposing flexor and extensor muscles in each rat's hind foot, which resulted in reversing the movement of the ankle, to see if the animals could learn to move the foot correctly, as was predicted by Weiss's functionalist view. He was surprised to find that the rats never adjusted, even after long hours of training.[10] For example, while climbing a ladder their foot went up when it should have gone down and vice versa. He had assumed new circuits would be established and normal function would return, but it turned out that motor neurons were not interchangeable. Next he tried the sensory system, transposing the skin nerves from one foot to another. Once again the rats continued to have false reference sensations: when the right foot was shocked, they would lift the left one; when the right foot had a sore, they would lick the left one.[11] Both their motor and sensory systems lacked plasticity. Unfortunately, Weiss had made a poor choice in picking the salamander to use as a model for the human in his experiments; regeneration of the nervous system is exhibited only by the lower vertebrates, that is, fishes, frogs, and salamanders. Sperry was returning to the idea that a type of chemotaxis regulated the growth and termination of nerve fibers, first proposed early in the twentieth century by one of the greatest neuroscientists of all time, Santiago Ramón y Cajal.

Sperry thought that the growth of nerve circuits was the result of a highly specific genetic coding for nerve contacts. He performed dozens of clever experiments to make his point. In one, he simply took a frog and surgically turned the eye upside down. Afterwards, when the frog was shown a fly, his tongue went for it in the opposite direction. Even after the eye had been in this position for months, the frog continued to search for it in the wrong direction. There was specificity to the system: it was not plastic and could not adapt. He then took a goldfish and cut parts of the retina. As the nerves regenerated, he watched where they would grow in the part of the midbrain that receives input from the eyes, the optic tectum. It turned out that they would grow very specifically. If they were growing from the back of the retina, they would grow to the front of the tectum, and if they were from the front of the retina, they grew to the back of the tectum. In other words, there was a specific location that they grew to, no matter their starting position. Sperry concluded that "Whenever central fiber systems were disconnected and transplanted or just scrambled by rough surgical section, regrowth always led to orderly functional recovery and under conditions that precluded re-educative adjustments."[12] A bit later in the 1960s, nerve growth was actually observed and photographed, revealing that the growing tip of the nerve continuously sends out several microfilaments, or feelers, that probe in all directions, elongating and retracting as they sense which direction to extend the growth of the nerve.[13] Sperry maintained that chemical factors determined which microfilament would dominate and set the course of growth. In his model for neuron growth, neurons grow out to find their connection in the brain by sending out little filopodia (slender cytoplasmic projections from the cell) to see which way to go—testing the waters so to speak—and because of a chemical gradient, they would find their way to a specific place.

This fundamental idea has led to the notion, still prevalent in neuroscience today, of neural specificity. Sperry's original model has been altered and changed with subtle adjustments and some tweaking, but his general model for neuronal growth remains. With neuron growth and connectivity under genetic control, the overall result of this mechanism of neuron growth is that the brain's organizational scheme throughout the vertebrate kingdom is generally the same. Leah Krubitzer, an evolutionary neurobiologist at the University of California–Davis, thinks that it is probable that there is a common genetic pattern for the cortex for all species determined by the same genes. She summarizes, "This would explain the persistence of a common plan of organization or a blueprint for development in every mammal examined, and the existence of vestigial sensory apparatus and cortical areas in mammals that do not appear to use a particular sensory system."[14] Some parts get pushed around a bit by different sizes and shapes of skulls and brains, but the relationships have the same overall plan.

Whereas Lashley's and Weiss's experiments seemed to show that different areas of the brain were undifferentiated and interchangeable, Sperry had shown that the opposite was true: Most cerebral networks are determined genetically by some chemical or physiochemical coding of pathways and connections. This is a hard-wired view in which the differentiation, migration, and axon guidance of nerve cells is under genetic control. But there was a problem with a pure nativist view that the mind possesses ideas that are only inborn and not derived from external sources. The limits on this idea had been foreshadowed by Hebb.

EXPERIENCE

About the same time as Sperry was fine-tuning his theory of nerve development in the early 1960s, a young British biologist, Peter Marler, became fascinated with songbirds. These birds learned their songs from their fathers. He had noticed, while doing botanical fieldwork, that songbirds of the same species had somewhat different songs (which he called dialects) in different locales. Looking at white-crowned sparrows, he found that young sparrows were eager and able to learn a range of sounds during a brief sensitive period from about 30 to 100 days old. He wondered if he could control what song they learned by what song they were exposed to. He isolated young birds during this sensitive period and exposed them to the songs of either their home dialect or an alien dialect. They learned the dialect that they were exposed to. So the dialect they learned was dependent upon their experience. Then he wondered if they could learn the slightly different song of a different species of sparrow if they were exposed to one. He tried alternating the training song with the song of a different sparrow species that was common in their native habitat, but they learned only the song of their own species.[15] So while the song dialect that they learned depended on the song that they were exposed to, the variations of the song that they were able to learn were very limited. There were preexisting neural constraints in what they were able to learn. These built-in constraints presented a problem for the blank-slaters, but did not surprise Niels Jerne.

SELECTION VERSUS INSTRUCTION

In the 1950s, Niels Jerne, the famous Swiss immunologist, rocked the world of immunology to its core. At the time, there was a nearly unanimous consensus among immunologists that antibody formation was equivalent to a learning process in which the antigen played an instructive role. Antigens are usually proteins or polysaccharides that make up parts of cell surfaces. These cells can be microbes, such as bacteria, viruses, or parasites, or nonmicrobial, such as pollen, egg white, or protein from transplanted organs, tissues, or on the surface of transfused blood cells. Jerne suggested that something quite different was happening. He suggested that instead of a specifically designed antibody being formed when an antigen presented itself, the body was born supplied with all the different types of antibodies that it was ever going to have: Antigens were merely molecules that were recognized or selected by one of these innate antibodies. No instruction was going on, just selection. The complexity is built into the immune system, it doesn't become more so over time. His ideas are the foundation for what is now known about antibody response and clonal selection theory (the cloning, that is, the multiplying, of white blood cells, aka lymphocytes, with receptors that bind to invading antigens). Most of these antibodies will never encounter a matching foreign antigen, but those that do are activated and produce many clones of themselves to bind and inactivate the invading antigen.

Jerne kept on shaking things up. He later suggested that if the immune system works on this selection process, then most likely other systems do, too, including the brain. Jerne wrote an article in 1967, entitled "Antibodies and Learning: Selection versus Instruction,"[16] on the importance of viewing the brain as responding to selection processes and not to instruction: The brain was not an undifferentiated mass that could learn anything, just as the immune system was not an undifferentiated system that could produce any type of antibody. He made the startling suggestion that learning may actually be the process of sorting through preexisting capacities that we innately possess to apply to a particular challenge facing us at a moment in time. In other words, these capacities are genetically determined neural networks specialized for particular kinds of learning. An oft-used example is that it is easy to learn to be afraid of snakes, while it is difficult to learn to be afraid of flowers. We have a built-in template that elicits a fear reaction when we detect certain types of motion, such as slithering in the grass, but no such innate reaction to flowers. Here, just as in the immune system, the idea is that complexity is built into the brain, along with the specificity we talked about above as exemplified by the white-crowned sparrow's song. The very important idea is that there is selection from preexisting capacity. But it also implies constraints. If the capacity is not built-in, it does not exist.

A famous example of selection at work in the world of population biology was observed in Darwin's original classroom, the Galápagos Islands. In 1977 a drought led to a crop failure of most of the seed-producing shrubs, and it resulted in a high mortality rate of adult medium ground finches. The ground finches had beaks of variable size. The finches ate a diet of seeds, and their livelihood involved their beaks. The finches with smaller beaks were unable to crack the woody Tribulus fruits and hard seeds that remained proportionally common during the time of drought, but the larger-beaked finches could. The scant supply of softer seeds was gobbled up quickly, leaving only the larger, tough seeds that could only be eaten by the birds with the larger beaks. The small-beaked finches perished, leaving the larger-beaked finches: selection from preexisting capacity. The following year, the offspring of the surviving birds tended to be bigger and had bigger beaks.[17]

The current view of the brain is not the brain as depicted by Lashley, Watson, and Weiss. Their model featured the brain as an undifferentiated mass ready to learn: Any brain could learn anything. For such a brain, it would be as easy to teach it to enjoy the fragrance of roses as the fragrance of rotten eggs; it would be just as easy to teach it to be afraid of flowers as to be afraid of snakes. I don't know about yours, but essence of rotten eggs wafting out from the kitchen is not going to impress my dinner guests, no matter how many times they come to dinner. Sperry challenged this conception and argued that the brain is built in a very specific way, genetically determined, and that we arrive from the baby factory mostly prewired. This explanation, however, while explaining most of the facts, didn't explain all the data that continued to pour in from ongoing research. It didn't completely explain Marler's songbirds.

ACTIVITY-DEPENDENT PROCESS

It turns out that, as usual in neuroscience, there was more to the story. Wun Sin, Kurt Kass, and their colleagues, studying neuron growth in the optical tectum of frog brains, found that by supplying a light stimulus, they could increase the growth rate and the number of branching projections, or dendritic spines, at the tip of the nerve cell. These dendritic spines conduct electrical stimulation from other nerve cells and are known collectively as the dendritic arbor. Thus, enhanced visual activity actually drove nerve growth.[18] Rather than growth solely affected by a type of genetically driven chemotaxis (where cells direct their movement toward certain chemicals) that Sperry proposed, the actual activity of the neuron, its experience, also drives its growth and the neuronal connections it subsequently forms. This is known as an *activity-dependent process*.

Annoyingly, it has recently been shown that Mom was right: I should have practiced piano more. In fact, practicing any motor skill does make perfect. Practice not only changes the efficacy of synapses,[19] but recently it has been shown[20] that the synaptic connections in the living mouse rapidly respond to motor skill training and permanently rewire. Training a month-old mouse to reach with its forelimb caused rapid (within an hour!) formation of dendritic spines. After training, the overall spine density returned to the original level by eliminating some of the old spines and stabilizing the new spines formed during learning. These same researchers also showed that different motor skills are encoded by different sets of synapses. The good news is that it may not be too late for me (or at least mice) to heed my mother's advice. Practice of new tasks also promotes dendritic spine formation in adulthood. The bad news is, I still would have to practice. Motor learning appears to be the result of actual synaptic reorganization, and the stabilized neuronal connections appear to be the foundation of durable motor memory.

Associative learning is another example of how experience can alter neural connectivity. If you have seen the movie *Seabiscuit,* you may remember when Seabiscuit was being retrained to start running at the sound of a bell. When the bell would sound, the horse would also get a thump on his backside with a riding crop that evoked the "flight" response, and he started to run. After several trials he would run with just the sound of the bell, and off he went to beat the East Coast champion, War Admiral.

Thus, while the overall connectivity pattern is under genetic control, outside stimuli from the environment and training also affect neuronal growth and connectivity. The current view of the brain is that its large-scale plan is genetic, but specific connections at the local level are activity-dependent and a function of epigenetic factors and experience: Both nature and nurture are important, as any observant parent or pet owner can report.

Preexisting Complexity

Human developmental psychology is overflowing with examples of what babies intuitively know about physics, biology, and psychology. For years, Elizabeth Spelke, at Harvard, and Rene Baillargeon, at the University of Illinois, have studied what babies know about physics. This is knowledge that adult humans take for granted and rarely wonder where it has come from. For instance, the coffee cup on your desk would normally not attract much of your visible interest. If, however, you were to suddenly see your coffee cup zoom up to the ceiling, it would then attract your interest in a big way, and you would stare at it. It would have defied gravity. You expect objects to conform to a set of rules, and if they don't, you stare at them. You would have stared at that cup even if you had never learned about gravity in school. The same thing applies to a baby. If his bottle were to suddenly fly up to the ceiling, he would stare at it.

By taking into account that babies will look longer at objects that aren't conforming to a set of rules, researchers have teased out what those rules are for a baby. Baillargeon placed a ball in front of three-and-a-half-month-old babies and then blocked it with a screen. She then secretly removed the ball. When the screen fell back and no ball was present, the babies were shocked. This is because they already seem to grasp the physics that mass can't pass through mass. By three and a half months old, babies expect objects to be permanent and to not disappear when blocked from view.[21] In a number of other experiments, they have shown that infants expect objects to be cohesive rather than to spontaneously splinter apart if tugged on. They also expect them to keep the same shape if they pass behind a screen and reemerge: A ball shouldn't turn into a teddy bear. They expect things to move along continuous paths and not to travel across gaps in space; and they make assumptions about partially hidden shapes: the visible half sphere when fully revealed should be a ball, it shouldn't have legs. They also expect an object not to move on its own unless something contacts it, and to be solid and not to pass through another object.[22] This is knowledge that is genetically determined and which we are born with. How do we know this isn't learned knowledge? Because babies everywhere know the same stuff at the same age no matter what they have been exposed to.

Preexisting complexity also seems built into the human visual system. At the level of human perception, many other automatic processes are also built in. For example, in the visual realm, what is there is not necessarily what you see. It has long been known that two squares with the same measured light intensity appear to be of different brightness when presented on two different backgrounds. A gray square on a darker background appears to be lighter than the same square on a lighter background.

The luminance of an object is basically determined by the light shining on it, the reflection from its surface, and the transmittance of the space (for instance, whether there is fog or a filter) between the observer and the object. The perception of an object's luminance is called brightness. There is not, however, a simple correspondence between an object's luminance and its perceived brightness. If there are changes in any of these three variables, the relative intensity of the light reaching the eye may or may not be different depending on the combinations of the variables. For example, look at the four walls of the room that you are sitting in. They may all be painted the same color but one may appear lighter than another, depending on how it is illuminated. One wall may appear bright white and another light gray, while the third may be dark gray. Come back later in the day when the light has changed, and so may have the brightness of the walls. Thus there is no fixed relationship between the source of a visual stimulus and the elements that combine to produce the stimulus; and no way for the visual system to figure out how these factors have combined to produce the luminance values in the image that reaches the retina.

Why would such a system evolve? Duke University researchers Dale Purves, Beau Lotto, and colleagues point out that successful behavior requires responses compatible with the origin of the stimulus, rather than the measureable properties of the stimulus; this can only be learned by past experience, both the individual's past and his evolutionary past.[23] For instance, learning the luminance of a ripe fruit hanging against a background of foliage would be more advantageous than its specific optical properties. In other words, they suggest that the visual circuitry and the resulting perception have been selected according to past successes of visually guided behavior. "If this idea is correct, then to the extent that the stimulus is consistent with similarly reflective target surfaces under the same illuminant, the targets will tend to appear similarly bright. However, insofar as the stimulus is consistent with the past experience of the visual system with differently reflective objects in different levels of illumination, the targets will tend to appear differently bright."[24] The point is we are not cognitively aware of this. Our visual perception system has evolved through a process of selection to have this complex automatic mechanism built in.

THE ROAD TO *HOMO SAPIENS*

Paleoanthropologists estimate that modern-day humans share with the chimpanzee a common ancestor that lived somewhere between five and seven million years ago. For some reason, often blamed on a change in weather that may have caused a change in the food supply, there was a split in our common line. After a few false starts and unsuccessful branchings, one line eventually evolved into the chimps (*Pan troglodytes*) and the other into *Homo sapiens*. Although we *Homo sapiens* are the only surviving hominid from this line, many came before us. The few fossil remains of these hominids provide us clues about how we have evolved.

Our First Bipedal Ancestor

One hominid fossil in particular caused quite a stir. In 1974 Donald Johanson shocked the anthropological world when he uncovered the first approximately four-million-year-old fossil remains of what became known as an *Australopithecus afarensis*. Nearly 40 percent of the skeleton was found, and portions of the pelvic bone revealed it to be the remains of a female: the now famous Lucy. The discovery of Lucy was not what was so shocking. What was shocking was she was fully bipedal but did not have a big brain. Up until that time, it was thought that our ancestors had evolved a big brain first and that big brain and its big ideas had led to bipedalism. A few years later, in 1980, Mary Leakey found fossil footprints of A. *afarensis* dating to 3.5 million years ago that had nearly identical shape, form, and weight distribution as our footprints, providing additional evidence of full-blown bipedalism before the evolution of a big brain. More recently, Tim White and colleagues have made another fascinating discovery. They have found several fossils, including an intermediate foot, of *Ardipithecus ramidus* dating to 4.4 million years ago.[25] With each fossil discovery, theorists are being sent back to the drawing board. Tim White and his international team now suggest that our last common ancestor with the chimp was less chimplike than has been generally thought and that the chimp itself has undergone more evolutionary changes since the divergence than has been previously appreciated.

One such theorist, the psychologist Leon Festinger, was curious about the origins of modern man, and wondered just exactly which of our ancestors could be recognized as the earliest human. He pointed out that bipedalism must have been a "nearly disastrous disadvantage,"[26] for it greatly reduced the speed of movement, both for running and for climbing. In addition, while a four-legged animal can still run quite well on three legs, a bipedal animal with an injured leg cannot. This disadvantage obviously made it more vulnerable to predators.

Becoming bipedal produced another disadvantage: The birth canal became smaller. A wider pelvis would have made bipedalism mechanically impossible. Embryonically, the skulls of primates form in plates that slide over the brain and do not coalesce until after birth. This allows the skull to remain pliable enough to fit through the birth canal, but also allows the brain to grow after birth. At birth, a human baby has a brain that is about three times larger than that of a baby chimp, but it is developmentally less advanced. Thus, in comparison with other apes, we are born one year prematurely, producing another disadvantage: Human babies are helpless and need to be cared for longer. After birth, however, there is a striking difference between the developing child's brain and that of the chimp. The child's brain continues to expand through the adolescent years and triples in size with all kinds of refinements and influences going on in this plastic period. It ends up weighing about 1300 grams. A chimp's brain, however, is nearly fully developed at birth and ends up with a weight of about 400 grams.

Bipedalism must have had some advantage that allowed our ancestors to survive and successfully reproduce. Festinger suggested the advantage these hominids had was not that they had two free appendages to use for things other than locomotion, but they had a brain that was now inventive enough to figure out what those possible other uses could be: "The arms and hands were not, and are not, specialized appendages as, for example, the legs of man are. An extraordinary variety of uses was invented for the arms and hands—and invented is the key word." Owen Lovejoy, speculating about the *Ardipithecus ramidus* fossils, suggests that those appendages were used by males to carry food to females in a food-for-sex swap, setting a whole constellation of physiological, behavioral, and social changes into play.[27] Festinger suggested that ingenuity and imitation drove brain evolution: "All the humans that lived about two and a half million years ago need not have had the idea of manufacturing tools with sharp edges. . . . If one individual, or a small group of them, invent a new process, others can, and do, imitate and learn." Most of what we humans do originated with just one human's smart idea, which we copy. Who was it that made that first cup of joe from those rather uninteresting-looking beans? That was someone else with a different brain from mine. Luckily, however, I didn't have to reinvent the wheel. I can use another guy's smart idea. Invention and imitation are ubiquitous in the human world, but are shockingly rare among our animal friends.

Reduced speed of movement and more predators, however, while a seeming disadvantage, may have been the grand instigator of many cognitive changes. That early inventive brain had to first solve the predator problem, the origin of the phrase "necessity is the mother of invention." Of the two ways to outfox a predator, one is to be bigger and faster—an unworkable option. The other is to live in large groups, increasing both surveillance and protection, but also making hunting and gathering more efficient. Many ideas have been batted around over the years trying to figure out what forces were driving the relentlessly enlarging hominid brain. It now seems to be boiling down to two factors that were driving the processes of natural and sexual selection: a diet that provided the added calories needed to feed the metabolically expensive bigger brain, and the social challenges originating from living in those larger groups needed for protection.

Does the fact that we have this larger brain account for the differences between us and other animals?

Holloway Takes On the Big Brain Idea

The idea that human capabilities are merely a function of a larger brain stems from Charles Darwin, who wrote "the difference between man and the higher animals, great as it is, is certainly one of degree and not of kind."[28] His champion and ally, neuroanatomist T. H. Huxley, denied humans had any unique brain features other than size.[29] This idea that the only difference between the brains of humans and our nearest primate relatives was one of size persisted unquestioned until the 1960s. Then Ralph Holloway, now a professor of anthropology at Columbia University, threw his hat in the ring. He suggested that evolutionary changes in cognitive capacity are the result of brain reorganization, not changes in size alone.[30] He writes, "I came to this conclusion before 1964 when I made a seminar presentation [. . . .] demonstrating that some human microcephalics with brain sizes that some gorillas might deride as diminutive were nevertheless able to talk. That meant to me that something in their brains was organized differently than in the great apes."[31] Finally in 1999, Todd Preuss and his colleagues were able to provide Holloway with some physical evidence. They were the first to find microscopic differences in brain organization between apes and humans.[32]

More support comes from evolutionary biologists Willem de Winter and Charles Oxnard. They have suggested that a brain part's size is related to its functional relationships with other brain parts. They ran a multivariate analysis (an analysis of more than one statistical variable at a time) using brain part ratios from 363 species and found that groups of species with more similar brain part ratios emerged based on similar lifestyles (locomotion, foraging, and diets), rather than on phylogenetic (evolutionary) relationships. For instance, New World insectivorous bats had brain part ratios more closely linked with Old World carnivorous bats, rather than with their phylogenetically closer relatives the New World fruit-eating bats. De Winter's and Oxnard's analysis revealed that the species within a lifestyle group had similar brain organizations: that the convergence and parallels in brain relationships are most likely associated with convergences and parallels in *lifestyles* that cut across phylogenetic groups.[33] Falling into a group unto themselves, humans, of the 363 species studied, were the only species to have a bipedal lifestyle. They found a highly significant 22-standard-deviation difference* between the brain organization of humans and chimpanzees. He concluded, "The nature of human brain organization is very different from that of chimpanzees, which are themselves scarcely different from the other great apes and not too different even from Old World monkeys."[34]

It isn't surprising that Darwin posited a change in degree only. Although all species are unique unto themselves, we are made up of many

---

* Standard deviation measures the spread of data around a mean value. If the standard deviation is large, then the variability is great. A normal distribution of data usually falls within 3 standard deviations greater or lesser than the mean.

of the same molecular and cellular building blocks and have evolved by the same principles of natural selection. Antecedents to all kinds of things we previously assumed were uniquely human have been observed. Yet Yale neuroanatomist Pasko Rakic, in an admonition that we will hear from others, reminds us "We are all seduced by the remarkable similarities in cortical organization within and between species, such that we forget that the differences are where we should look for the evolutionary progress that has led to the ascent of our cognitive abilities."[35]

This disagreement about how the human brain differs from other animals, and indeed how the brains of other animals differ from one another, whether it is one of quantity versus quality continues, but the evidence for a truly qualitative difference, a difference in kind, is far more compelling. The great psychologist David Premack, who tried teaching language to chimpanzees for years, is in the same camp as Rakic: "The demonstration of a similarity between an animal and a human ability should automatically trigger the next question: What is the dissimilarity? This question will prevent mistaking similarity for equivalence."[36]

One of the major dissimilarities that Premack emphasizes is that the abilities of other animals do not generalize: Each species has an extremely limited set of abilities and these abilities are adaptations restricted to a single goal: Scrub jays plan for the single goal of future food but not for other things, nor do they teach or make tools in the wild. Crows make tools in the wild but only for obtaining food, and they do not plan or teach. Meerkats don't plan or make tools in the wild, but they teach their young one thing only: how to eat poisonous scorpions without being stung. None can take their skill and adapt it across many domains. The meerkat teaches its young how to safely eat scorpions. Humans, on the other hand, teach everything to their young, and what is taught usually generalizes to other skills. In short, teaching and learning have been generalized.

As with other animals, the core constituents of human abilities also evolved as specific adaptations, and humans possess an unrivaled number of highly refined abilities that evolved in this fashion. The combination of these abilities has given rise to additional abilities for solving general problems, leading to domain-general abilities that are uniquely human. The result is an explosion of ability and realization of the human condition. Modern neuroanatomists are quick to point out that as you climb the primate scale to humans, it is not that additional skills are simply being added on as once was hypothesized,* but the whole brain is getting rearranged throughout. We are still left with this thorny little problem: What is going on in the brain to produce this magnificent ability that humans have, how did it come about, and how do you capture it? Fortunately for job security and today's graduate students, the mystery is alive and well, but some of the secrets are being revealed, which we will now explore.

PHYSICAL DIFFERENCES EXHIBITED BY THE HUMAN BRAIN

All these assaults on the big brain theory have sent researchers to the microscope with more advanced techniques for counting cells and of staining cells to bring out their details. And now, an unbridgeable crack in the foundation of the big brain theory is opening in front of our eyes.

Bigger Isn't the Answer to Better

A few problems shadowed the big brain theory even before we were presented with microscopic differences in 1999. Neanderthals had big-

---

* The triune brain model hypothesized by Paul Maclean. In this model, the structure of the brain is based on its evolutionary development, and is made up of three layers, the earliest reptilian layer, overlaid by the limbic system, with the newest layer, the neocortex, encircling the other two. Basically what he suggested was that as we evolved, we kept adding brain layers, just as you would add a car onto a train. I call it the train theory of evolution.

ger brains than humans, but never expressed the scope of abilities that we possess. Over the course of history, there has been a decrease in the brain size of *Homo sapiens*. My attention was drawn to this question while studying patients with intractable epilepsy, who have undergone split-brain surgery. In this procedure, the large tract of nerves that connects the two hemispheres, the corpus callosum, is severed to prevent the spread of electrical impulses. Their isolated left brain, however, which receives no input from the right hemisphere (in essence losing half its size), remains just as intelligent as a whole brain. If brain quantity is so important, you would think that there would be an effect on problem solving and hypothesizing when half the brain is lost, but there is not.

Playing the game of championing neuron numbers is now up against another problem. Just as Mark Twain claimed that "The reports of my death have been greatly exaggerated," so too have the claims that humans have a bigger brain than is expected for an ape of our size. In 2009, using a new technique to count neurons, Frederico Azevedo and his coworkers[37] found that in terms of numbers of neuronal and non-neuronal cells, the human brain is a proportionately scaled-up primate brain: It is what is expected for a primate of our size and does not possess relatively more neurons.* They also found that the ratio between nonneuronal brain cells and neurons in human brain structures is similar to those found in other primates, and the numbers of cells match those expected for a primate of human proportions. In fact, instead of humans being the outliers among the primates with a larger-than-expected brain for body size, they concluded that embarrassingly for orangutans and gorillas, they are outliers with a larger-than-expected body for their brain size.

---

* They determined that the adult male human brain contains on average 86 billion neurons and 85 billion nonneuronal cells and while the cerebral cortex is 82 percent of the brain's mass, it possesses only 19 percent of the brain's neurons. The majority of the neurons, 72 percent, were found in the cerebellum, which makes up 10 percent of the mass of the brain.

The human brain has on average 86 billion neurons, but 69 billion of them are located in the cerebellum, that small structure at the back of the brain that helps refine motor control. The entire cortex, the area that we think is responsible for human thought and culture, has only 17 billion, and the rest of the brain has a little less than one billion. The frontal lobes and prefrontal cortex—the part of the human brain that is involved with memory and planning, cognitive flexibility, abstract thinking, initiating appropriate behavior and inhibiting inappropriate behavior, learning rules, and picking out relevant information perceived through the senses—have vastly fewer neurons than the number in the visual areas, the other sensory areas, and the motor areas of the cortex. What is larger in the frontal lobes than the rest of the brain, however, is the arborization of the neurons, that branching of the dendritic tips of the neurons with the resulting possibility of increased connections.

Now the brain anatomists have their work spelled out for them. If the number of human neurons is just scaled up from a chimp's brain, then their connectivity patterns or the neurons themselves must be different.

Connectivity Changes

When brain size increases, what is increasing is the number of neurons, their connections, and the space between the neurons. The finding that the human cerebral cortex volume is 2.75 times larger than in chimpanzees, but has only 1.25 times more neurons[38] intimates that a good deal of the increased mass is due to the space between cell bodies and what that space is filled with. The space, known as neuropil, is filled with the stuff that connections are made of: axons, dendrites, and synapses. In general, the larger the area, the better connected it is,[39] as more neurons connect to more and more other neurons. As the brain is scaled up, however, if every neuron were to connect with every other neuron, the increased volume of connections and the increased length of connections stretching across the increasing size, would slow down nerve-signal processing speed, and the overall benefit would be trivial.[40] What happens is not every neuron is connected to every other neuron. There is a fall in percentage of connectedness. At some point, as absolute brain size and total neuron number increases, the proportional connectivity decreases and the internal structure changes as the connectivity pattern changes. In order to add new function, the decrease in proportional connectivity forces the brain to specialize. Small local circuits, made of an interconnected group of neurons, are created to perform specific processing jobs and become automatic. The result of their processing is passed on to another part of the brain, but all the computations that were used to arrive at the result are not. So as we discussed with the visual perception problem, the result of the processing—the judgment that the gray square appears lighter or darker—is passed on, but the processing steps that arrived at that conclusion are not.

The past forty years of research have shown that the human brain has billions of neurons organized into local, specialized circuits for specific functions, known as modules. For instance, in the human brain, an example of different circuits running in parallel and processing different inputs was demonstrated by a neuroimaging study done by Mark Raichle, Steve Petersen, and Mike Posner. One part of the brain reacts when you hear words, another particular part of the brain reacts to seeing words, still another area reacts while speaking words, and they can all be going at the same time.[41] James Ringo, who realized this need for larger brains to have a decreased proportional connectivity resulting in more specialized networks, also pointed out that this explains the problem with Karl Lashley's rats and their equipotential brains: Their small brains had not formed specialized circuits that are characteristic of larger brains. Now add to this discussion Todd Preuss's comment, "The discovery of cortical diversity could not be more inconvenient. For neuroscientists, the fact of diversity means that broad generalizations about cortical organization based on studies of a few 'model' species, such as rats and rhesus macaques, are built on weak foundations."[42]

Throughout mammalian evolution, as brain size has increased, the size of the evolutionarily youngest part of the brain, the neocortex, has increased disproportionately. The six-layered neocortex is made up of neuronal cells, Monsieur Poirot's "little gray cells," and sits like a large, wrinkled napkin on top of the cerebral cortex. The neocortex is responsible for sensory perception, generation of motor commands, spatial reasoning, conscious and abstract thought, language, and imagination. This increase in size is regulated by the timing of neurogenesis, which of course is under control of DNA. With a longer developmental period comes more cell divisions, which result in a larger brain. The outermost layers, the supragranular layers (II and III), mature last[43] and project primarily to other locations within the cortex.[44] About these layers, Jeff Hutsler, from our lab, made an important observation: In primates there is a greater proportional increase of layer II/III neurons than in other mammals. They make up 46 percent of the cortical thickness of a primate, 36 percent of carnivores, and 19 percent of rodents.[45] This layer is thicker because of its dense network of connections between cortical locations. Many researchers think that this layer and its connections participate heavily in higher cognitive functions by linking motor, sensory, and association areas. The different thicknesses of these layers in different species may imply a corresponding unequal degree of connectivity,[46] which could play a role in the cognitive and behavior differences of various species.[47] Increases in neocortex size would allow recrafting of local cortical circuits and an increased number of connections.

As primate brains have increased in size, however, the corpus callosum, the large neuronal fiber tract that transmits information between the two hemispheres of the brain, has become proportionately smaller.[48] Increased brain size is thus associated with reduced interhemispheric communication. As we have evolved toward the human condition the two hemispheres have become less hooked up. Meanwhile, the amount of connectivity within each hemisphere, the number of local circuits, has increased, resulting in more local processing. While many circuits are symmetrically duplicated in both sides of the brain (for instance, the right brain has circuits that for the most part control the movements of the left side of the body and the left has circuits that control the right side of the body), there are many circuits that exist in only one hemisphere. Local circuits that have lateralized, meaning that they are present in only one of the two hemispheres, are rampant in the human brain. In the past few years, we have learned of neuroanatomical asymmetries in many animal species, but humans appear to have lateralized circuits present to a far greater degree.[49]

Some of the framework for our lateralization must have been already present in out last common ancestor with the chimp. For example, my colleagues Charles Hamilton and Betty Vermeire, while they were investigating the ability of the macaque monkey to perceive faces, discovered right-hemisphere superiority for the detection of monkey faces,[50] just as humans exhibit for human faces. Others have found that both humans and chimps have asymmetrical hippocampi (the structure that regulates learning and consolidation of spatial memory, mood, appetite, and sleep), the right is larger than the left.[51] The hominid line, however, has undergone further lateralization changes. In the search for asymmetries between the other primates and humans, areas involved with language have been the most studied and many asymmetries have been found in them. For instance, the planum temporale, a component of Wernicke's area, the cortical area associated with language input, is larger on the left side than the right side of humans, chimps, and rhesus monkeys, but it is microscopically unique in the left hemisphere of humans: The cortical minicolumns[*] are wider, and the spaces between them are greater. The resulting different neuronal

---

* The individual neurons within the six-layered sheet of the neocortex line up with those in the sheets above and below to form columns (aka microcolumns or minicolumns) of cells that cross the sheets perpendicularly.

structure may indicate that there is a more elaborate and less redundant way that information is processed in the left hemisphere. It may also indicate that there is another component in this space that is still unknown. Asymmetries in the cortical structure of the posterior language region and Broca's area, involved with speech comprehension and production, also exist indicating that there have been changes in connectivity that are responsible for this unique ability.[52]

Early in split-brain research, we came across another startling anatomical difference. In the brains of chimpanzees and rhesus monkeys, the anterior commissure, a fiber tract that connects the middle and inferior temporal gyri of the two hemispheres, is involved with the transfer of *visual* information.[53] We have learned from more recent studies on split-brain patients, however, that the anterior commissure does not transfer visual information in humans, but does transfer olfactory and auditory information: same structure, different function. Another marked difference is the major visual pathway, which projects from the retina of the eye to the primary visual cortex in the occipital lobe (the back of the brain) in both monkeys and humans: Monkeys with damage to their visual cortex can still see objects in space, and discriminate color, luminance, orientation, and patterns.[54] Humans with the same lesions, however, cannot perform these tasks and are blind. Such differences in capacities of corresponding brain tracts, underlines the fact that species differences between similar structures are at work, and that, once again, we need to be wary of cross-species comparisons.

A new imaging technique, diffusion tensor imaging, can actually map nerve fibers. The way that the human brain is organized on the local level is now obtainable, seeable, detectable, and quantifiable. More evidence for changes in neural connectivity patterns have been identified using this technology. For instance, the white matter fiber tract that in humans is involved with language, the arcuate fasciculus, has a completely different organization in the monkey, the chimp, and the human.[55]

Different Types of Neurons

A few years ago I wondered if anyone thought there were differences in nerve cells from one species to another or if they were all the same. I asked several leading neuroscientists: If you were recording electrical impulses from a slice of the hippocampus in a dish, and you were not told if the slice came from a mouse, a monkey, or a human, would you be able to tell the difference? At the time, most of the responses would have accorded with this answer that I received: A cell is a cell is a cell. It's a universal unit of processing that only differs in size between the bee and the human. If you appropriately size a mouse, monkey, or human neuronal cell you won't be able to see the difference even if you had Pythia to help you. But now, there is a heretical view that has been coming on in the last ten years: All neurons are not alike, and some types of neurons may be found only in specific species. Moreover, a given type of neuron may exhibit unique properties in a given species.

The first evidence of microscopic differences in the neuronal arrangement was found in 1999 by neuroanatomist Todd Preuss and his colleagues and was located in the primary visual cortex in the occipital lobe. They found that the neurons of cortical layer 4A differ both architecturally and biochemically in humans from the other primates. The layer these neurons make up is part of the system that relays information about object recognition from the retina through the visual cortex in the occipital lobe to the temporal lobe. In the human brain, they form a complex, meshlike pattern, differing from the simple vertical pattern found in the other primates. This was unexpected, for, as Preuss put it, "In visual neuroscience, the proposition that there are no important differences between macaques and humans is something close to an article of faith."[56] Preuss conjectured that this evolutionary change in neuronal arrangement may be responsible for humans' superior ability to detect objects against a background.

The ramifications of this finding involve the fact that most of our understanding of the structure and function of the visual system is from studies primarily done on macaque monkeys. As already noted, this finding and others that demonstrate cortical diversity are, as Preuss put it, at the very least, inconvenient. The generalizations that neuroscientists have made about neuronal architecture, cerebral organization, connections, and resulting function have been based on the studies of a few species, namely macaques and rats. Just how faulty a foundation this is has yet to be determined and is not confined to the visual system.

Even the basic building block of the brain, the pyramidal neuron, so called because its cell body is shaped like a Hershey's Kiss, has come under scrutiny. In 2003, after decades of comparative neuroscience research extolling the commonalities of pyramidal neurons across species, Australian Guy Elston reaffirmed and brought to our attention the original insights of Santiago Ramón y Cajal. Just as David Premack was concerned about similarities being interpreted as equivalences when comparing behavior between species, so Elston mourns that, among comparative neuroscientists studying mammalian cerebral cortices, "unfortunately, 'similar' was interpreted by many to imply 'same.'" This resulted in the widespread acceptance that the cortex is uniform and is made up of the same basic repeated unit, and that this basic unit was the same in different species.[57] This didn't make any sense to Elston, who wondered, "if circuitry in prefrontal cortex, the region of the brain often implicated in cognitive processing, is the same as that in any other cortical region, how could it perform such a complex function as human mentation?" Nor did it make sense to Cajal, who a hundred years ago concluded, after a lifetime of research, that the brain was not built of the same repeated circuit.

Elston and others have found that the branching patterns and numbers of the basal dendrites in the prefrontal cortex's (PFC) pyramidal neurons are greater than in other areas of the cortex. Those dendrites provide each pyramidal neuron in the PFC greater connectivity than in other parts of the brain. Potentially, this means that individual neurons in the PFC receive a larger number of more diverse inputs over a bigger region of cortex than their counterparts in other parts of the brain. Indeed, differences in the pyramidal cell are not confined to regional differences only. He and his colleagues have found, without the aid of Pythia, that the pyramidal cell is characterized by marked differences in structure among primate species.[58]

There is also evidence that nerve cells do not respond the same in all species. During neurosurgery, when a tumor is removed, some normal neuronal cells are also removed. Yale University neurobiologist Gordon Shepherd has found that when he puts these human cells in tissue culture and records from them, and then does the same with guinea pig neuronal cells, the way that the dendrites respond is different in the two species.[59]

Still Other Types of Neurons

In the early 1990s Esther Nimchinsky and coworkers at the Mount Sinai School of Medicine decided to restudy a rather rare and forgotten neuron first described in 1925 by the neurologist Constantin von Economo.[60] The long, thin von Economo neuron (VEN) differs from the more squatty pyramidal neuron in that it is about four times larger, and while both have a single apical (situated on the top) dendrite, it has only a single basal dendrite, as opposed to the many-branched pyramidal neuron. They also are only found in specific areas of the brain involved with cognition—the anterior cingulate cortex and the frontoinsular cortex—and have recently been identified in the dorsolateral prefrontal cortex of humans[61] and elephants. Among primates, VENs are found only in humans and the great apes,[62] and humans have the most, both in absolute and relative numbers. They found that while the average number in the great apes was 6,950, the adult human has 193,000 cells, a four-year-old human child has 184,000, and a newborn has 28,200. Because of their location, structure, biochemistry, and diseases of the nervous system that they are involved with, neuroscientist John Allman at the California Institute of Technology and his colleagues[63] propose that they are part of the neural circuitry involved in social awareness and may participate in fast, intuitive social decision making. In the hominid line, VENS appear to have arisen in the common ancestor of the great apes about 15 million years ago. Interestingly, the only other mammals in which they have been found are also large-brained social animals: elephants,[64] some types of whales,[65] and most recently in dolphins.[66] These neurons have risen independently and are an example of convergent evolution, which is the acquisition of the same biological trait in evolutionarily unrelated lineages. While the von Economo neuron is not unique to humans, the extent to which we possess them is.

Undetermined as yet is if the predecessor cells in the thirty-one- to fifty-one-day-old human embryo, discovered and named in 2006 by Irina Bystron and her colleagues, are uniquely human.[67] As their name implies, these are the first neurons that form in the cerebral cortex. No equivalent of these cells has been found in any other species.

WE JUST AREN'T ALL WIRED THE SAME

With this mounting evidence of physical anatomical differences, differences in connectivity, and differences in cell type, I think that we can say that the brains of humans and the brains of other animals appear to differ in how they are organized, which, when we truly come to understand it, will help us understand what makes us so different.

So here we are, born with this wildly developing brain under tremendous genetic control, with refinements being made by epigenetic factors (nongenetic factors that cause the organism's genes to behave differently) and activity-dependent learning. It is a brain with structured, not random, complexity, with automatic processing, with a particular skill set with constraints, and with a generalized capacity that has all evolved through natural selection. We will see in the chapters ahead that we have myriad cognitive abilities that are separated and spatially represented in different parts of the brain, each with different neural networks and systems. We also have systems running simultaneously, in parallel, distributed throughout the brain. This means that our brains have multiple control systems, not just one. From this brain comes our personal narrative, not from some outside mental forces compelling the brain.

Yet many mysteries lie ahead. We are going to try to understand why we humans, who have no problem accepting that our body's housekeeping mechanisms, such as breathing, are a result of our brain's activity, are so resistant to the idea that the mind is embodied in the brain. Another conundrum we'll look at is why the idea that we are born with a complex brain, not a blank one that can be easily changed, has been seemingly hard to swallow. We will see that the way our brain functions, and our beliefs and feelings about how it functions, impact not only the ideas of downward causation, consciousness, and free will, but also our behavior.

But what does this mean to any of us? As Bob Dylan might ask, how does it feel to understand how we got here? How does it feel to wonder if we are freely choosing moral agents or to wonder how it all works? Does a person who believes that the human mind, its thoughts, and resulting actions are determined, actually feel any different than anyone else? And, after a couple more chapters, how will it feel to grasp why we feel psychologically unified and in control even though we may not be? Ahhh . . . not much different. I'm not really having an existential crisis, if that's what you're worried about. No doubt you will still feel pretty much in control of your brain, in charge, and calling all the shots. You will still feel that someone, you, is in there making the decisions and pulling the levers. This is the homuncular problem we can't seem to shake: The idea that a person, a little man, a spirit, *someone* is in charge. Even those of us who know all the data, who know that it has got to work some other way, we still have this overwhelming sense of being at the controls.

Stay tuned.

Chapter Two

THE PARALLEL AND DISTRIBUTED BRAIN

DO YOU REMEMBER THE TELLING SCENE IN THE MOVIE *MEN in Black*, when a corpse is undergoing an autopsy? The face popped open only to reveal the underlying brain machinery, and right there was an extraterrestrial-looking homunculus pulling levers to make it all work. It was the "I," the "self," the phenomenal center and take-charge thing we all think we have. Hollywood captured it perfectly, and we all believe in it even though we may understand that that is not at all how it works. Instead, we understand that we are stuck with these automatic brains, these vastly parallel and distributed systems that don't seem to have a boss, much like the Internet does not have a boss. So much of us comes from the factory all wired up and ready to go. Think about the wallaby, for example. The last ninety-five hundred years have been Easy Street for the Tamar wallabies that live on Kangaroo Island off the coast of Australia. They have lived there without a single predator to worry them all those years. They have never even seen one. So why then, when presented with stuffed models of predatory animals such as a cat, fox, or the now-extinct animal that had been their historical predator, do they stop foraging and become vigilant, but they don't when presented with a model of a nonpredatory animal? From their own experience, they shouldn't even know that there are such things as animals they should be wary of.

Like the wallaby we have thousands, if not millions, of wired-in predilections for various actions and choices. I don't know about wallaby minds, but we humans think we are making all our decisions to act consciously and willfully. We all feel we are wonderfully unified, coherent mental machines and that our underlying brain structure must somehow reflect this overpowering sense we all possess. It doesn't. Again, no central command center keeps all other brain systems hopping to the instructions of a five-star general. The brain has *millions* of local processors making important decisions. It is a highly specialized system with critical networks distributed throughout the 1,300 grams of tissue. There is no one boss in the brain. You are certainly not the boss of the brain. Have you ever succeeded in telling your brain to shut up already and go to sleep?

It has taken hundreds of years to accumulate the knowledge we currently have about the organization of the human brain. It has been a rocky road as well. While the story unfolded, however, a persistent uneasiness remained about the knowledge. How can stuff be localized in the brain in so many ways and still seemingly function as an integrated whole? The story begins a long time ago.

LOCALIZED BRAIN FUNCTIONS?

The first hints came from anatomy, and the modern understanding of human brain anatomy stemmed from studies done by the English physician Thomas Willis, he of the circle of Willis* fame, in the seventeenth century. He was the first to describe the longitudinal fibers of the corpus callosum and several other structures. A little over a hundred years

---

* The vascular structure at the base of the brain.

later, in 1796, Franz Joseph Gall, an Austrian physician, came up with the idea that different parts of the brain produced different mental functions which resulted in a person's talents, traits, and dispositions. He even suggested that morality and intelligence were innate. Although these were good ideas, they were based on a faulty premise that was not grounded in good science. His premise was that the brain was composed of different organs, and that each was responsible for a mental process that resulted in a specific trait or faculty. If a particular faculty were more highly developed, its corresponding organ would enlarge and could be felt by pressing on the surface of the skull. From this idea he suggested that one could then examine the skull and diagnose that individual's particular abilities and character. This became known as phrenology.

Gall had another good idea: He moved to Paris. As the story goes, however, he displeased Napoleon Bonaparte by not attributing to his skull certain noble characteristics that the future emperor was sure he possessed. Obviously Gall was no politician. When he applied to the Academy of Science of Paris, Napoleon ordered the academy to get some scientific evidence for his conjectures, so the academy asked the physiologist Marie-Jean-Pierre Flourens to see if he could come up with any concrete findings that would back up this theory.

At that time there were three methods of inquiry that Flourens could tackle: (1) surgically destroying specific parts of animal brains and observing the results; (2) stimulating parts of animal brains with electrical pulses and seeing what happened; (3) or studying neurological patients clinically and performing autopsies on them after their deaths. Flourens became quite taken with the notion that specific locations in the brain performed specific processes (cerebral localization), and went with option one to investigate this idea. Studying rabbit and pigeon brains, he became the first to show that yes, certain parts of the brain were responsible for certain functions: When he removed the cerebral hemispheres, that was the end of perception, motor ability, and judgment; without the cerebellum the animals became uncoordinated and lost their equilibrium; and when he cut out the brain stem—well, you know what happened—they died. He could not, however, find any areas for advanced abilities like memory or cognition, just as psychologist Karl Lashley, who we read about in the last chapter, would later observe studying rat brains. He concluded that these functions were more diffusely scattered throughout the brain. Examining skulls to determine character and intelligence did not stand up to the rigors of science, and slid into the hands of charlatans. Unfortunately, Gall's good idea, that there was localization of cerebral function, got tossed out with the bad. His other good idea, moving to Paris, has been well accepted.

Not too many years later, however, evidence relevant to Gall's idea started trickling in from clinical studies. In 1836 another Frenchman, Marc Dax, a neurologist in Montpellier, sent a report to the Academy of Sciences about three patients, noting the coincidence that each had speech disturbances and similar left-hemisphere lesions found at autopsy. A report from the provinces, however, didn't get much air time in Paris. It wasn't until nearly thirty years later that anyone took much notice of this observation that speech could be disrupted by a lesion to one hemisphere only. This happened in 1861, when a well-known Parisian physician, Paul Broca, published his autopsy on a patient who had been nicknamed Tan. Tan had developed aphasia and was so named because *tan* was the only word he could utter. Broca found that Tan had a syphilitic lesion in his left hemisphere, in the inferior frontal lobe. He went on to study several more patients with aphasia, all with lesions in the same area. This region, later called the speech center, is also known as Broca's area. Meanwhile, German physician Carl Wernicke was finding patients with lesions in an area of their temporal lobe, who could hear words and sounds just fine, but could not understand them. The hunt for sites in the brain for specific abilities was off to the races.

Hughlings Jackson, a British neurologist, confirmed Broca's findings, but he is a part of this story in his own right. His wife suffered generalized seizures, which he was able to observe very closely. He noticed that they always started in a specific part of her body and progressed systematically in a pattern that did not vary. This suggested to Jackson that specific areas of the brain controlled the motor movements of different parts of the body and gave rise to his theory that motor activity originated from and was localized in the cerebral cortex. He also wielded an ophthalmoscope that had been invented a few years earlier by Hermann von Helmholtz, the German physician and physicist. This instrument allows physicians to peer into the nether reaches behind the eye. Jackson thought it important for neurologists to study the eye, and why this is so will become apparent as we travel further on. From these early clinical observations that were followed up with autopsy findings, it was looking more and more as if Gall had been on the right track about cerebral localization of functions.

THE GREAT WORLD OF THE UNCONSCIOUS

Localization was not the only idea about cerebral functioning that had been simmering. Fictional writing, from Shakespeare's *Othello* to Jane Austen's *Emma,* was implying that much was going on in the nonconscious brain department. While Sigmund Freud tends to get the credit for the buried iceberg of nonconscious processes, he was not the originator of the idea, but the trumpet. Many, notably the philosopher Arthur Schopenhauer, from whom many of Freud's ideas sprung, preceded him in emphasizing the importance of the unconscious and as did later, the English Victorian version of a Renaissance man, Francis Galton. Galton wore many hats. He was an anthropologist, tropical explorer (Southwest Africa), geographer, sociologist, geneticist, statistician, inventor, meteorologist, and was also considered the father of psychometry, which is the development of both instruments and techniques for measuring intelligence, knowledge, personality traits, and so forth. In the journal *Brain*,* he painted a picture of the mind as if it were a house set upon a "complex system of drains and gas- and water-pipes . . . which are usually hidden out of sight, and of whose existence, so long as they acted well, we had never troubled ourselves." In the conclusion to this paper he wrote: "Perhaps the strongest of the impressions left by these experiments regards the multifariousness of the work done by the mind in a state of half-unconsciousness, and the valid reason they afford for believing in the existence of still deeper strata of mental operations, sunk wholly below the level of consciousness, which may account for such mental phenomena as cannot otherwise be explained."[1] Galton, unlike Freud, was interested in basing his theories on concrete findings and statistical methods. He added to the researchers' armamentarium the statistical concepts of correlation, standard deviations, and regression to the mean, and was also the first to use surveys and questionnaires. Galton was also interested in heredity (no wonder—his cousin was Charles Darwin). Galton was the first to use the term *nature versus nurture* and to do studies on twins to tease out the varying influences.[†]

So emerging into the twentieth century, the ideas of localized brain functions and nonconscious processes were being batted around, but as we saw in the last chapter, these ideas suffered a detour early in the twentieth century with the wide acceptance of behaviorism and the equipotential brain theory. The theory of an equipotential brain, however, had always faced a serious challenge from clinical medicine. This began with Dax's observation of the correlation of a lesion in a specific part of the brain with a specific result in a variety of people. The equipotential brain theory never could explain this or many of the other seemingly mysterious cases from neurology. Once scientists understood, however, that the brain has distributed and specialized net-

---

* Cofounded by Hughlings Jackson.

† Galton, a man of many firsts, also devised the classification system used to identify fingerprints and figured out the probabilities that two people would share the same fingerprint.

works, some of these clinical mysteries could be solved. Even before the advent of modern brain imaging and EEG techniques, studying the deficits of patients with lesions allowed a type of reverse engineering and provided insights into how the brain enables cognitive states.

HELP FROM PATIENTS

Neuroscientists owe much to the many clinical patients who have generously participated in our research. Studying clinical patients with X-rays and early scanning devices began to reveal that all sorts of unusual behaviors were caused by lesions in specific locations. For instance, a lesion in one specific area of the parietal lobe can produce the odd syndrome of *reduplicative paramnesia*, a delusional belief that a place has been duplicated or exists in more than one spot at the same time, or has been moved to a different location. I had a patient who, although she was being examined in my office at New York Hospital, claimed we were in her home in Freeport, Maine. I started with the question "So, where are you?" She replied, "I am in Freeport, Maine. I know you don't believe it. Dr. Posner told me this morning when he came to see me that I was in Memorial Sloan-Kettering Hospital and that when the residents come on rounds to say that to them. Well, that is fine, but I know I am in my house on Main Street in Freeport, Maine!" I asked, "Well, if you are in Freeport and in your house, how come there are elevators outside the door here?" She calmly responded, "Doctor, do you know how much it cost me to have those put in?"

As we proceed toward the front of the brain, a lesion in the lateral frontal lobes produces deficits in sequencing behavior, leaving one unable to plan or multitask. Orbital frontal lesions, located right above the eye sockets, may interrupt the emotional pathways that give feedback to monitor cognitive states and may be associated with a loss of the ability to judge right and wrong. There can be a decreased ability to inhibit behavior, leading to more impulsive, obsessive-compulsive, aggressive, and/or violent actions and higher-order cognitive dysfunctions. And in the left temporal lobe, a lesion in Wernicke's area produces Wernicke's aphasia, where the affected person may have no comprehension of either written or spoken language, and although he or she may speak fluently with a natural language rhythm, it's gibberish. So from clinical medicine, we can see that specific parts of the brain are involved with particular aspects of cognitive activity.

FUNCTIONAL MODULES

Indeed, today it seems that localized brain function is very much more specific than even Gall may have considered. Some patients have lesions in the temporal lobe that leave them very poor at recognizing animals but not man-made artifacts and vice versa.[2] A lesion in one spot leaves you unable to tell a Jack Russell from a badger (not that there is much difference), and with damage in another spot, the toaster is unrecognizable. There are even people with certain brain lesions who specifically cannot recognize fruit. Harvard researchers Alfonso Caramazza and Jennifer Shelton claim that the brain has specific knowledge systems (modules) for animate and inanimate categories that have distinct neural mechanisms. These domain-specific knowledge systems aren't actually the knowledge itself, but systems that make you pay attention to particular aspects of situations, and by doing so, increase your survival chances. For example, there may be quite specific detectors for certain classes of predatory animals such as snakes and big cats.[3] A stable set of *visual* clues may be encoded in the brain that make you pay attention to certain aspects of biological motion, such as slithering in the case of snakes or sharp teeth, forward-facing eyes, body size, and shape in the case of big cats, which are used as input to identify them.[4] You don't have innate knowledge that a tiger is a tiger, but you may have innate knowledge that when you see a large animal with forward-facing eyes and sharp teeth that stalks, it is a predator and you automatically become wary; similarly, you automatically get a little shot of adrenaline and veer away from the slithering movement in the grass.

This domain specificity for predators is not limited to humans, of course. Richard Cross and colleagues at the University of California–Davis studied some squirrels that had been raised in isolation and had never seen a snake in their lives. When exposed to snakes for the first time they avoided them, but did not avoid other novel objects: The squirrels had an innate wariness of snakes. In fact these researchers have been able to document that it takes ten thousand years of snake-free living for this snake template to disappear in populations.[5] And this explains our Kangaroo Island wallabies. They were reacting to some visual cue that the stuffed predators exhibited, not to any behavior or odor. Thus, highly specific modules exist, in this case for identification, that do not require prior experience or social context to work. These mechanisms are innate and hard-wired; some of these we share with other animals; some animals have mechanisms that we don't have; and some are uniquely human.

SPLITTING THE BRAIN

Starting in 1961, there was a new opportunity to study the brain at work, with patients who had had their cerebral hemispheres divided, the so-called split-brain patients. In the late 1950s, Roger Sperry's lab at Caltech was studying the effects of dividing the corpus callosum (CC) in monkeys and cats,[6] and was developing new methods to test for these effects. They had found that if they taught one hemisphere a task in animals with an intact CC, the skill would transfer to the other hemisphere, but that if the CC were divided, it did not. The divided brains had divided perception and learning. Big effects were being found, and the question presented itself, could similar effects be found in the human? There was a great deal of skepticism, for a few reasons. Although many neurological cases reported in the late nineteenth century described specific impairments with focal lesions in the CC, these findings were the victims of Lashley's equipotential cerebral cortex theory and had been ignored, swept under the carpet, and literally forgotten about for many years. More seeming evidence for the skeptics was that children who were born without a CC showed no ill effects.* The final big reason was that in a series of twenty-six patients who each had had their corpus callosum cut (known as a commissurotomy) to treat intractable epilepsy at the University of Rochester in the 1940s, no neurological or psychological consequences had been observed by a gifted young neurologist, Andrew Akelaitis, who had tested them.[7] The patients all felt just fine after their surgery and they themselves noticed no differences. Karl Lashley had seized on these findings to push his idea of mass action and equipotentiality of the cerebral cortex; discrete circuits of the brain were not important, he claimed—only cortical mass. He suggested that the function of the corpus callosum was simply to hold the hemispheres together.

In the summer between my junior and senior year at Dartmouth College, I landed in Roger Sperry's lab at Caltech as an undergraduate summer fellow because I was interested in the nerve regeneration studies I wrote about in the last chapter. The lab, however, was now focused on the corpus callosum, so I spent the summer trying to anesthetize the half brain of a rabbit and decided basic research was the life for me. I was captivated by the question of what was happening to humans after callosal surgery. Because the lab was finding dramatically altered brain function in the cats, monkeys, and chimps with callosal sections, I was convinced there had to be some effect on humans. During my senior year, I came up with the plan of retesting Akelaitis's Rochester patients during my spring break and designed a different method of testing them. Armed with my first grant of $200 from the Hitchcock Foundation at Dartmouth Medical School to cover a rental car and hotel room, I drove to Rochester. My rental car was full of borrowed taschistoscopes

---

* Later it was concluded that they had developed compensatory pathways.

(pre–computer age devices that display images on a screen for a specific amount of time) from the Dartmouth psychology department for the testing that had been set up. While I was waiting to begin, however, the testing was canceled, and I was left empty-handed and disappointed. However, my curiosity was unabated; I was determined to return to the vibrant atmosphere of Caltech for graduate school, which came about the following summer.

To begin my graduate studies, a new opportunity presented itself. Dr. Joseph Bogen, a neurosurgery resident at the White Memorial Hospital in Los Angeles, and his attending physician, Philip Vogel, had a patient whom Bogen, after critically reviewing the medical literature, thought would benefit from a split-brain procedure; the patient agreed. For the previous ten years, this patient, a robust and charming man, WJ, had been suffering two grand mal seizures a week, each of which took him a day to recover from. Obviously this had an enormous impact on his life, and he was ready to risk the surgery. Already armed with the testing procedures that I had designed at Dartmouth, I was assigned to test WJ, both before and after his surgery.[8] His surgery was a great success, and he was electrified by the facts that he felt no different and his grand mal seizures were completely resolved. I was electrified, too, by what I discovered about WJ's brain function and have been fascinated with the results from this patient and those who followed ever since.

The surgical procedure to cut the CC was performed after all other treatments for intractable epilepsy had been tried. William Van Wagenen, a Rochester, New York, neurosurgeon, performed the procedure for the first time in 1940, following the observation that one of his patients with severe seizures got relief after developing a tumor in his corpus callosum.[9] It was thought that if the connection between the two sides of the brain were cut, then the electrical impulses causing the seizures wouldn't spread from one side of the brain to the other, and a generalized convulsion would be prevented. Splitting the brain in half, however, is a big deal. The great fear was what the side effects of the surgery might be. Would it create a split personality with two brains in one head? In fact the treatment was very successful. Patients' seizure activity decreased on average 60–70 percent, some were totally free of seizures altogether, and they all felt just fine: no split personality, no split consciousness.[10] Most seemed completely unaware of any changes in their mental processes. They appeared completely normal. This was great, but puzzling nonetheless.

The procedure for fully dividing the hemispheres includes cutting the two fiber pathways that connect the hemispheres: the anterior commissure and the corpus callosum. Not all the connectivity between the two hemispheres is severed, however. Both hemispheres are still connected to a common brain stem, which supports similar arousal levels, so conveniently, both sides sleep and wake at the same time.[11] The subcortical pathways remain intact and both sides receive much of the same sensory information from the body's nerves relating to the five senses and proprioceptive information from the sensory nerves in the muscle, joints, and tendons about the body's position in space. At the time, we didn't know that both hemispheres can initiate eye movements and that there also appears to be only one integrated spatial attention system, a set of processes that allows the selection of some stimuli over others, which continues to be unifocal after the brain has been split. Thus, attention cannot be distributed to two spatially disparate locations:[12] unfortunately, contrary to what most modern drivers are quite sure they are capable of, the right brain can't watch the traffic while the left brain is reading text messaging. We have since also learned that emotional stimuli presented to one hemisphere will still affect the judgment of the other hemisphere. We did know initially, as we learned earlier from the studies of Dax and Broca, that our language areas are located in the left hemisphere (exceptions are in a few left-handed people).

When WJ was tested before his surgery, he could name objects presented to either visual field or objects placed in either of his hands. He could understand any command and carry it out with either hand, that is to say, he was normal. When he returned for testing after surgery, WJ felt just fine, and like the patients from Rochester, noticed no changes, except that he was no longer having seizures. I had devised a testing procedure, which, unlike the one used by Akelaitis, took advantage of the anatomy of the human visual system. In humans, the optic nerves from each eye meet at what is called the optic chasm. Here, each nerve splits in half, and the medial half (the inside track) of each crosses the optic chasm into the opposite side of the brain and the lateral half (that on the outside) stays on the same side. The parts of both eyes that attend to the right visual field send information to the left hemisphere and information from the left visual field goes to and is processed by the right hemisphere. In the animal experiments, this information did not cross over from one disconnected hemisphere to the other. Only the right side of the brain had access to information from the left visual field and vice versa. Because the visual system is set up in this manner, I could feed information to one half of the animals' brains only.

The day arrived for WJ's first test after his surgery. What would we find? Things, at first, progressed as expected. We expected that because his speech center was located in his left hemisphere, he would be able to name objects seen by his left hemisphere. Accordingly, he could easily name objects presented to his left hemisphere. Thus, when we flashed a picture of a spoon in the right visual field and then asked, "Did you see anything?" He quickly replied, "A spoon." Then came the initial critical test: What would happen when these objects were presented to his right hemisphere from the left visual field? Akelaitis's work had suggested that the corpus callosum played no essential role in the interhemispheric integration of information. Thus it could be predicted that WJ would be able to describe the object normally. The animal studies being done at Cal Tech, however, suggested otherwise, and that was where I was putting my money. We flashed a picture to his left hemisphere and I asked, "Did you see anything?"

If you are not engaged in scientific research, you may better understand the electricity of the moment if you think of a roulette wheel spinning around with a couple years' wages riding on red. You would be hoping that the ball would land on red, anticipation mounting as the wheel begins to slow, with your livelihood and hours of work invested in the outcome. I was hoping that my experimental design would reveal something as of yet unknown, and my anticipation grew as the time approached to flash a picture to the right hemisphere. What would happen? Adrenaline was pumping through my body, my heart was bouncing around like a football at Dartmouth when Bob Blackman was the coach. While the findings are old hat now and fodder for cocktail party discussions, there is no describing my amazement when WJ said, "No, I didn't see anything." Not only could he no longer verbally describe, using his left hemisphere, an object presented to his freshly disconnected right hemisphere, but he did not know that it was there at all. The experiment that I had designed as an undergraduate and was able to do as a graduate student had revealed a startling discovery! Christopher Columbus could not have felt any more excited on spotting land than I felt at that moment.

Initially, it seemed, he was blind to stimuli presented to his left visual field. On further investigation, however, this was not the case. I had another trick up my sleeve to ferret out whether the right hemisphere was receiving any visual information. While both hemispheres can guide the facial and upper arm proximal muscles, the separate hemispheres have control over the hand's distal muscles. Thus, the left hemisphere controls the right hand and the right hemisphere controls the left hand.[13] If the hands are kept out of sight, then the left brain has no idea what the left hand is up to, and vice versa. I devised an experiment in which WJ could respond using a Morse code key with his left hand (controlled by his right hemisphere) before giving a verbal response (controlled by his left hemisphere). I flashed a light to his right hemisphere; he responded by pressing the key with his left hand when it flashed, but stated that he saw nothing! His right hemisphere was not blind to the stimuli, it saw the flash just fine and could report it using the Morse code key. The only reason for WJ denying the light flash had to be that there was a total disruption of the transfer of information between the two hemispheres!

It turned out that any visual, tactile, proprioceptive, auditory, or olfactory information that was presented to one hemisphere was processed in that half of the brain alone, without any awareness on the part of the other half. The left half did not know what the right half was processing, and vice versa. I found that a split-brain patient's left hemisphere and language center has no access to the information that is being fed to the right brain. We were being presented a completely new opportunity: to study the presence of an ability in one hemisphere separated from the other hemisphere, not a deficit caused by a lesion.

In later experiments with other patients, we put assorted objects within reach of the left hand but blocked from view. A picture of one of the objects was flashed to the right hemisphere, and the left hand felt among the objects and was able to select the one that had been pictured. When asked, however, "Did you see anything?" or "What is in your left hand?" the patient denied seeing the picture and could not describe what was in his left hand. In another scenario we flashed the picture of a bicycle to the right hemisphere and asked the patient if he had seen anything. Once again he replied in the negative, but his left hand drew a picture of a bike.

It soon became apparent that the right hemisphere was superior at visual spatial skills. While the left hand, under right-hemisphere control, could easily put together a series of colored blocks to match a pattern in a picture flashed to the right hemisphere, the right hand, when the picture was flashed to the left hemisphere, took forever to solve the puzzle. In fact, one patient had to sit on his left hand to prevent it from coming up and trying to solve the problem. The left hand could copy and draw three-dimensional pictures, but the right hand, that one that so easily can write a letter, could not draw a cube. The right hemisphere turned out to be specialized for such tasks as recognizing upright faces, focusing attention, and making perceptual distinctions. The left hemisphere was the intellectual. It specialized in language, speech, and intelligent behavior. After commissurotomy, the verbal IQ of a patient is unchanged,[14] as is his problem-solving capacity. There may be some deficits in free-recall capacity and in other performance measures, but isolating essentially half of the cortex from the dominant left hemisphere caused no major change in cognitive functions. The left remains unchanged from its preoperative capacity, yet the largely disconnected, same-size right hemisphere is seriously impoverished in cognitive tasks. It was becoming apparent that the right hemisphere had its own rich mental life, quite different from that of the left.

We already knew from the study of neurological patients that the brain had two completely different neuronal pathways for generating spontaneous facial expressions and voluntary ones. Only the dominant left hemisphere could generate voluntary facial expressions.[15] In patients who have a particular lesion in their right hemisphere that disrupts communication between the hemispheres, only the right side of the face responds when asked to smile, and the left side remains immobile.* If the same patient is told a joke and spontaneously smiles, however, his facial muscles respond normally bilaterally, because a different pathway is used that doesn't require communication between the hemispheres. The exact opposite is true with Parkinson patients who have damage in their extrapyramidal systems, the part of the motor system that is involved with the coordination of movements. They are unable to have spontaneous expressions, but can voluntarily control their facial muscles. In our split-brain experiments, we figured that if we gave a command to the left hemisphere of a patient, then the right side of the face should respond first, and this is exactly what happened. When the left hemisphere of a split-brain patient sees the command to smile or frown, the right side of the face responds about 180 milliseconds before the left side responds; the time lag is due to the right hemisphere's having to get the somatic feedback through subcortical pathways.

---

* The left hemisphere predominantly controls the facial muscles on the right, and right hemisphere controls those on the left.

All of these findings led to a picture of many specializations distributed around the brain. But another conclusion seemed to follow our studies: With the observation that each hemisphere could possess information outside the realm of awareness of the other half-brain, it suggested that the surgery had induced a state of double consciousness.

DOUBLE CONSCIOUSNESS?

Not everyone was excited by these findings. While riding up in the elevator at Rockefeller University, George Miller introduced me to the great American psychologist William Estes and said, "You know Mike, he is the guy that discovered the split-brain phenomenon in humans?" and Estes responded, "Great, now we have two systems we don't understand!" It appeared that split-brain surgery produced two separate conscious hemispheres and, at the time, we thought there were two conscious systems: mind left and mind right.

In 1968 Roger Sperry wrote: "One of the more general and also more interesting and striking features of this syndrome may be summarized as an apparent doubling in most of the realms of conscious awareness. Instead of the normally unified single stream of consciousness, these patients behave in many ways as if they have two independent streams of conscious awareness, one in each hemisphere, each of which is cut off from and out of contact with the mental experiences of the other. In other words, each hemisphere seems to have its own separate and private sensations; its own perceptions; its own concepts; and its own impulses to act, with related volitional, cognitive, and learning experiences."[16]

Four years later I went overboard and added even more to this: "Over the past ten years we have collected evidence that, following midline section of the cerebrum, common normal conscious unity is disrupted, leaving the split-brain patient with two minds (at least), mind left and mind right. They coexist as two completely conscious entities, in the same manner as conjoined twins are two completely separate persons."[17]

This posed the problem of whether each consciousness had its own protagonist: Were there then two selves? Were there also two free wills? Why aren't the two halves of the brain conflicting over which half is in charge? Is one half in charge? Were the two selves of the brain trapped in a body that could only be at one place at one time? Which half decided where the body would be? WHY WHY WHY was there this apparent feeling of unity? Was consciousness and the sense of self actually located in one half of the brain?

WHAT IS CONSCIOUSNESS?

It was turning into a theoretical nightmare! And not only that, we were batting around the term *consciousness* and didn't really even know what it meant. No one had bothered to look it up. Years later, I decided to, and this is what I found in the 1989 *International Dictionary of Psychology*. The definition, written by psychologist Stuart Sutherland, was entertaining, if not edifying:

> CONSCIOUSNESS: The having of perceptions, thoughts, and feelings: awareness. The term is impossible to define except in terms that are unintelligible without a grasp of what consciousness means. Consciousness is a fascinating but elusive phenomenon; it is impossible to specify what it is, what it does, or why it evolved. Nothing worth reading has been written about it.[18]

That last bit was a relief to know, because more than eighteen thousand articles had been written about it the last time I did a Medline search, and Sutherland just told me not to bother reading them. You know you are treading on thin ice dealing with a topic when professionals are nervous about discussing it, and everybody else seems to think they understand it or has an opinion about it—something like explaining sex to your kids. At least if you are a physicist, the guy on the street isn't acting like he has string theory wired. The trouble with consciousness is that it has a mystique about it; we somehow want to treat it differently than, say, something like memory or instinct, which are also rather nebulous. We have not yet seen a physical instance in the brain of either of those but we have been able to slowly chip away at them, so I see no problem in tackling consciousness without having an exact definition. Neuroscientists are not alone with such problems. Researchers at the Santa Fe Institute recently told me that their current concept of the gene bears a weak resemblance to the original conception.

So while during the 1970s we were stuck with the idea that the split-brain patient was left with two conscious systems, Sir John Eccles and Donald MacKay were having none of it. Eccles, in his Gifford lectures in 1979 argued that the right hemisphere had a limited kind of self-consciousness, but not enough to bestow personhood, which resided in the left hemisphere. Donald MacKay was not satisfied with the idea, either, and commented in his Gifford lecture, "But I would say that the idea that you can create two individuals merely by splitting the organizing system at the level of the corpus callosum which links the cerebral hemispheres is unwarranted by any of the evidence so far. . . . It is also in a very important sense implausible."[19].

Well, science marches on, and we have left the idea of dichotomous mental systems in the dust, although, annoyingly, it still looms large in the popular press. With more patients to test, different testing methods, fancier equipment and brain scanners, much more data, the benefit of our own cerebral flexibility, and more smart people asking questions and designing experiments, we have moved toward the idea of a plethora of systems, some within a hemisphere and some distributed across hemispheres. We no longer think of the brain as being organized into two conscious systems at all but into multiple dynamic mental systems.

DICHOTOMOUS BRAIN THEORY
BITES THE DUST

The theory began to crack apart when we started to test the cognitive abilities of the right hemisphere and realized that the two hemispheres are not coequal. We had come to know the left hemisphere to be this whiz kid who could talk and understand language, while the right didn't talk and had a very limited understanding of language. So we began to give simple, first-grade-type conceptual tests to the right hemisphere using pictures and simple words it could understand. For example, when we flashed the word *pan* to the right hemisphere, the left hand would point to a pan. Next, we flashed the word *water,* and the left hand pointed to water. So far, so good: the right hemisphere could read the words and relate the words to the pictures. When we flashed the two together, however, the left hand could not put them together into the concept of water in a pan, and pointed to the empty pan picture. This water/pan task was quickly solved by the left hemisphere. It turns out that the right hemisphere is poor at making inferences. We tried presenting a problem only using pictures, such as flashing a picture of a match to the right hemisphere, followed by a picture of a woodpile, and then asking it to pick out one of six pictures that reflected the causal relationship. It could not pick out the picture of the burning woodpile. Even when using more visual-spatial stimuli, such as when we presented a form in the shape of a U and then asked which of a series of shapes would turn the U into a square, the right hemisphere was dismal at solving the puzzle. The left hemisphere, however, easily solved this problem. This difference was still present when some of our patients actually began to speak out of their right hemispheres and develop quite an extensive vocabulary: The right hemisphere still was unable to draw inferences.

This led us to the obvious conclusion that the conscious experience of the two hemispheres was very different. Among other things, one lived in a world where it could draw inferences, and the other did not.

The right hemisphere lives a literal life. When asked to decide whether various items appeared in a series of items previously shown to it, the right hemisphere is able to correctly identify items it saw previously and to reject new items. "Yes, there was the plastic spoon, the pencil, the eraser, and the apple." The left hemisphere, however, tends to falsely recognize new items when they are similar to previously presented items, presumably because they fit into the schema it has constructed.[20] "Yep, they are all there: the spoon [but we substituted a silver one for a plastic one], the pencil [although this one is mechanical and the other was not], the eraser [though it is gray and not pink], and the apple." As a consequence of not being able to draw inferences, the right hemisphere is limited by what it can have feelings about. A box of candy presented to the right hemisphere is a box of candy. The left hemisphere can infer all sorts of things from this gift.

If we had had Marcellus* around in our lab, perhaps he would have said, "There is something rotten in the state of dichotomous brain theory!" and we would have been forced to agree. Our findings gradually indicated to us that both halves of the brain had specializations, but each half of the brain was not equally conscious, that is, it was not conscious of the same things, and not equally capable of performing tasks. This was rotten enough for dichotomous brain theory, but absolutely stinking for the existing concepts about the unity of consciousness. Back to the drawing board with the question, Where is this conscious experience coming from? Does the information get processed and then channeled through one kind of conscious activation center that makes subjective experiences aware to you and me, or is it organized differently? The scales were tipping toward a different type of organization; a modular organization with multiple subsystems. We began to doubt that a single mechanism existed that enables conscious experience, but rather were heading toward the idea that conscious experience is the feeling engendered by multiple modules, each of which has specialized

* Thanks to William Shakespeare.

capacities. Since we were finding specialized capacities in all different regions of the brain and since we had seen that conscious experience was closely associated with the part of the cortex involved with a capacity, we came to understand that consciousness is distributed everywhere across the brain. Such an idea was directly contrary to that of John Eccles, who had championed the left hemisphere as the site of consciousness.

The essential observation that allows me to make this point is this: Right after split-brain surgery, when you ask a patient, "How are you?" The answer is "Fine." Then you ask," Do you notice anything different?" and the reply is "No." How could this be? You must remember that as the patient is looking at you, he cannot describe anything in the left part of his visual field. The left hemisphere, which is telling you that all is fine, cannot see half of what is in front of him and is not concerned about it. To compensate for this when not under testing conditions, split-brain patients will unconsciously move their heads to input visual information to both hemispheres. If you woke up from most other types of surgery and couldn't see anything in your left visual field, you would certainly be complaining about it, "Ahh, doc, I can't see anything on the left—what's up with that?" These patients never comment on this. Even after years of frequent testing, when asked if they know why they are being tested, they have no sense that they are special, no sense that anything is different about them or their brain. Their left brain does not miss their right brain or any of its functions. This has led us to realize that in order to be conscious about a particular part of space, the part of the cortex that processes that part of space is involved. If it is not functioning, then that part of space no longer exists for that brain or that person. If you are talking out of your left hemisphere, and I am asking you about your awareness of things in the left visual field, that processing is over in the disconnected right hemisphere and that hemisphere is conscious about it, but your left hemisphere is not. That area simply does not exist for the left hemisphere.

It doesn't miss what it doesn't have processing for, just like you don't miss some random person that you have never heard of.

This started us thinking that maybe consciousness is really a local phenomenon, and it is due to local processes associated with a particular sensory moment in left space or right space. This idea has allowed us to explain some of the previously inexplicable behaviors encountered in neurological patients.

Why do some people, who suddenly become blind in a large portion of their visual field complain about—are conscious of—it ("Hey, I can't see anything on my left side, what's going on?") and others don't say a word about—aren't conscious of—their sudden visual loss? The complainer's lesion is somewhere along his optic nerve, which carries information about vision to the visual cortex, the part of the brain that processes this information. If no information is coming in to a portion of his visual cortex, he is left with a blind spot and complains. The noncomplainer, however, has a lesion in the visual associative cortex (the part of the cortex associated with advanced stages of visual information processing that produces the visual experience) itself and not the optic nerve. The lesion also produces the very same blind spot, but the patient does not usually complain. Just like our split-brain patient does not complain. Why not? The visual cortex is the part of the brain that represents, or assembles the pictures from, the visual world. Each part of the visual field has a corresponding area in the visual cortex. So, for instance, there is an area that ordinarily asks, "What is going on to the left of visual center?" With a lesion on the optic nerve, this brain area is functioning; when it cannot get any information from the nerve, it puts up a squawk—"something is wrong, I am not getting any input!" When that very area of the associative visual cortex has a lesion, however, the patient's brain no longer has an area responsible for processing what is going on in that part of the visual field; for that patient that part of the visual field ceases to exist consciously; there is no squawk at all. The patient with the central lesion does not have a complaint, because the part of the brain that might complain has been incapacitated, and no other part takes over. The logical conclusion to these observations is that phenomenal consciousness, that feeling you have about being conscious of some perception, is generated by local processes that are uniquely involved with a specific activity.

I am suggesting that the brain has all kinds of local consciousness systems, a constellation of them, which are enabling consciousness. Although the feelings of consciousness appear to be unified to you, they are given form by these vastly separate systems. Whichever notion you happened to be conscious of at a particular moment is the one that comes bubbling up, the one that becomes dominant. It's a dog-eat-dog world going on in your brain with different systems competing to make it to the surface to win the prize of conscious recognition.

For instance, a few years after her surgery, one of our split-brain patients developed the ability to speak simple words out of her right hemisphere. This presents an interesting scenario, because it becomes a bit of a challenge to know which hemisphere is talking when she is speaking. In one interview she described her experience of looking at pictures of objects that were being flashed up on a screen in her different visual fields, "On this side [pointing to a picture on the left of the screen, flashed to her right hemisphere] I see the picture, I see everything more clearly; on my right side I feel more confident, in a way, with my answer." From previous testing, we knew that the right hemisphere was better at all kinds of perceptual judgments, so we knew that the statement about seeing more clearly was coming from her right hemisphere; and her confident speech center in her left hemisphere made the other. She put these two stories together, one from each hemisphere, but to the listener, it sounds like a completely unified statement coming from one unified system. We know intellectually, however, that it is information coming from two separate systems being woven together by our minds listening to her.

HOW DOES IT WORK?

How did we become so decentralized and end up with all these multiple systems? The answer harks back to what we touched upon in the last chapter in discussing the changes in connectivity patterns in big brains. With larger brains, more neurons, and increasing network size, proportional connectivity decreases. The number of neurons that each neuron is connected to remains about the same: The neuron does not connect up with more neurons as the total number rises for a few practical and neuroeconomical reasons. One is that if each neuron were connected to every other one, our brains would be gigantic. In fact two computational neuroscientists, Mark Nelson and James Bower, figured out that if our brains were fully connected and were the shape of a sphere, they would have to be 20 kilometers in diameter![21] Talk about having a big head. The metabolic costs would also be too great, with our brains constantly yelling "Feed me!" Currently our brains expend about 20 percent of the energy our bodies consume.[22] Imagine how much energy it would take to run a brain that was 20 kilometers across! (At least it would solve the obesity problem.) With long axons connecting neurons in distant parts of the brain, the processing speed would slow down, making synchronizing activity difficult. It would also require increased dendrite size in order to increase the number of synapses, and this would alter the electrical properties of the neuron, because the branching of the dendrites influences how it integrates electrical input from other neurons. No, our neurons could not feasibly all connect to each other; another solution was employed by our evolving brain.

Neurobiologist Georg Striedter, taking into account what is currently known about comparative neuroanatomy and connectivity in mammals, suggests that certain neuronal wiring "laws" apply to the evolutionary development of the large human brain.[23]

- *Decreased connectivity with increasing network size:* By maintaining absolute connectivity, not proportional connec-

tivity, large brains actually became more sparsely interconnected, but they had two tricks up their sleeve:

- *Minimizing connection lengths:* They maintained local
  connectivity using the shortest of connections.[24] Thus, less
  room was taken up with axons traveling back and forth, less
  energy was required to maintain the lines, and signaling was
  faster because it traveled over short distances. This set the
  stage for local networks to divide up and specialize, forming
  multiple clusters of processing modules. With all this
  separate processing, however, different parts of the brain
  must still exchange information and therefore, . . .

- Not all connections are minimized, but *some very long
  connections between distant sites are retained.* Primate brains
  have developed a "small-world architecture": many short, fast,
  local connections (a high degree of local connectivity), with a
  few long-distance ones to communicate their processing (a
  small number of steps to connect any two).[25] This design
  allows both a high degree of efficient local processing (modu-
  larity), and at the same time, quick communication to the
  global network. It is common to many complex systems,
  including human social relations.[26]

Our decentralization was the outcome of having a large brain and
the neuroeconomies which allowed it to function: less dense connections forced the brain to specialize, create local circuits, and automate.
The end result is thousands of modules, each doing their own thing.

Our conscious awareness is the mere tip of the iceberg of nonconscious processing. Below our level of awareness is the very busy nonconscious brain hard at work. Not hard for us to imagine are the
housekeeping jobs the brain constantly juggles to keep homeostatic
mechanisms up and running, such as our heart beating, our lungs
breathing, and our temperature just right. Less easy to imagine, but

being discovered left and right over the past fifty years, are the myriads of nonconscious processes smoothly putt-putting along. Think about it. To begin with there are all the automatic visual and other sensory processing we have talked about. In addition, our minds are always being unconsciously biased by positive and negative priming processes, and influenced by category identification processes. In our social world, coalitionary bonding processes, cheater detection processes, and even moral judgment processes (to name only a few) are cranking away below our conscious mechanisms. With increasingly sophisticated testing methods, the number and diversity of identified processes is only going to multiply.

OUR BRAIN'S JOB DESCRIPTION

What we always must keep in mind is that our brains, hence all these processes, have been sculpted by evolution to enable us to make better decisions that increase our reproductive success. Our brain's job description is to get its genes into the next generation. Years of split-brain research have made clear to us that the brain is not an all-purpose computing device, but a device made up of an enormous number of serially wired specialty circuits, all running in parallel and distributed across the brain to make those better decisions.[27] This network allows all sorts of simultaneous nonconscious processing to go on[28] and is what enables you to do things such as drive a car. You are simultaneously keeping in mind your route, judging distances between your car and those around you, your speed, when to brake, when to speed up, when to clutch and shift gears, remembering and following the traffic laws, and singing along with Bob Dylan on the radio, all at the same time. Pretty impressive!

Germane to our current discussion, however, is that while hierarchical processing takes place within the modules, it is looking like there is no hierarchy among the modules.* All these modules are not reporting to a department head, it is a free-for-all, self-organizing system. This is not the network that Gifford lecturer and neuroscientist Donald MacKay envisioned. He thought that conscious agency was the outcome of a central supervisory activity: "Conscious experience does not have its origin in any one of the participating brain nuclei, but in the positive feedback chain-mesh that is set up when the evaluative system becomes it own evaluator."

WHO OR WHAT IS IN CHARGE?

Yet we are still confronted with the question of why do we feel so unified and in control? We don't feel like there is a pack of snarling dogs in our brains. And why, for those who suffer from schizophrenia, does it feel as if someone else is in control of their actions or thoughts? Your friends at the cocktail party with no knowledge of psychology or neuroscience are fascinated or disbelieving if told about these nonconscious processes, only because they aren't apparent to the individual's personal experience. It's all very counterintuitive to us humans, with our strong sense of being unified into one self and feeling in control of our actions. Even among ourselves, we neuroscientists are having a hard time dispelling the idea of a homunculus, some central processor, calling the shots in our brains, such as Donald MacKay's proposal that we had a supervisory system overseeing our intentions and behavior that made adjustments to our environment. We may not actually say the "H" word, but use euphemisms such as "executive function" or "top-down processing." How can a system work without a *head* honcho

---

* Except in the sensory system. See: Bassett, D. S., Bullmore, E., Verchinski, B. A., Mattay, V. S., Weinberger, D. R., Meyer-Lindenberg, A. (2008). Hierarchical organization of human cortical networks in health and schizophrenia. *Journal of Neuroscience*, 28(37), 9239–9248.

and why does it feel like there is one? The answer to the first question may be that our brain functions as a complex system.

COMPLEX SYSTEMS

A complex system is composed of many different systems that interact and produce emergent properties that are greater than the sum of their parts and cannot be reduced to the properties of the constituent parts. The classic example that is easily understandable is traffic. If you look at car parts, you won't be able to predict a traffic pattern. You cannot predict it by looking at the next higher state of organization, the car, either. It is from the interaction of all the cars, their drivers, society and its laws, weather, roads, random animals, time, space, and who knows what else that traffic emerges.

In the past, it was thought that the reason such systems were complex was that not enough was known about them and that once all the variables were identified and understood, they would be completely predictable. Such a view is fully deterministic. Over the years, however, experimental data and theories are questioning such a conclusion. In fact, it is becoming accepted that complexity itself is rooted in the laws of physics, and we will discuss this further in chapter four. The study of complex systems is in itself complex and interdisciplinary, including not just physicists and mathematicians, but economists, molecular biologists up to population biologists, computer scientists, socialists, psychologists, and engineers.

Examples of complex systems are popping up all over the place: weather and climate in general, the spread of infectious disease, ecosystems, the Internet, and the human brain. Ironically for psychology in its quest to fully understand behavior, the signature phenomenon of a complex system "is the multiplicity of possible outcomes, endowing it with the capacity to choose, to explore and to adapt."[29] The implications of the idea that the human brain is a complex system has repercussions for discussions about free will, neuroscience and the law, and determinism, some of which we will discuss in later chapters.

Relevant to our current question about feeling unified and in control is an important point that Northwestern University's physicist Luis Amaral and chemical engineer Julio Ottino make: "The common characteristic of all complex systems is that they display organization without any *external* organizing principle being applied."[30] That means no head honcho, no homunculus.

All you have to think about is the Google ad auction to realize you can have a system that looks like someone is in charge, but no one is. It is run on algorithms. The ad auction has three selfish parties to please, the advertiser who wants to sell a product, thus needs a relevant ad; the user who wants relevant ads so he doesn't waste time; and Google, which wants satisfied advertisers and users to return for more business. Every time a user makes a query on Google, Google runs an auction for clicks. Advertisers have to pay only when they get a click. How this works is the advertisers provide a list of keywords, ads, and bids for how much they will pay when a person clicks their ad; however, the advertiser does not pay what he bids, he pays the bid of the advertiser that is below him in rank; that way he pays the minimum amount that is necessary to maintain ranking position. The Google user enters a search query, and Google compiles a list of ads whose keywords match the query. Google wants to be sure that the ads shown to users have a high quality. Quality is judged on three components. The most important is the click-through rate. Thus every time that a user clicks an ad, he votes on it. The second component is relevancy. Google looks to see how well the key words and context of an ad match up to the search query. It only uses relevant ads, saving shoppers from irrelevant ads by preventing ads from paying their way on to a search unrelated to their product. The third component is the advertiser's landing page quality, which should be relevant, easily navigable, and transparent. Ad rank is determined by the bid multiplied by the page quality. The beauty of the design is that the selfish motives of each party are harnessed and, voilà! As Google's chief economist points out, the most productive interaction results.[31] The system, while appearing to be run by a single controller, runs without one, by an algorithm.

Why do we feel so unified? We have discovered something in the left brain, another module that takes all the input into the brain and builds the narrative. We call this the interpreter module, and that is the topic of the next chapter.

Chapter Three

THE INTERPRETER

EVEN THOUGH WE KNOW THAT THE ORGANIZATION OF THE brain is made up of a gazillion decision centers, that neural activities going on at one level of organization are inexplicable at another level, and that as with the Internet, there seems to be no boss, the puzzle for humans remains. The lingering conviction that we humans have a "self" making all the decisions about our actions is not dampened. It is a powerful and overwhelming illusion that is almost impossible to shake. In fact, there is little or no reason to shake it, for it has served us well. There is, however, a reason to try and understand how it all comes about. Once we understand why we feel in charge, even though we know we live with a slight tape delay on what our brains are doing, we will understand why and how we make errors of thought and perception. In the next chapter, we will also be able to see where to look in our human life space for how personal responsibility comes to be and that it is alive and well in our reductionist world.

CONSCIOUSNESS: THE SLOW ROAD

When I was a kid, I spent a lot of time in the desert of Southern California—out in the desert scrub and dry bunchgrass, surrounded by purple mountains, creosote bush, coyotes, and rattlesnakes, where my parents owned some acreage. The reason I am still here today is because I have nonconscious processes that were honed by evolution. In particular, that snake template that I referred to in the last chapter. I jumped out of the way of many a rattlesnake, but that is not all. I also jumped out of the way of grass that rustled in the wind. I jumped, that is, before I was consciously aware that it was the wind that rustled the grass, rather than a rattler. If I had only my conscious processes to depend on, I probably would have jumped less, but would have been bitten on more than one occasion. Conscious processes are slow, as are what we consider to be conscious decisions.

As a person is walking, the sensory inputs from the visual and auditory systems go to the thalamus, a type of relay station. Then the impulses are sent to the processing areas in the cortex and then relayed to the frontal cortex. There they are integrated with other higher mental processes and perhaps the information makes it into the stream of consciousness, which is when a person becomes consciously aware of the information (there is a snake!). In the case of the rattler, memory then kicks in the information that rattlesnakes are poisonous and what the consequences of a rattlesnake bite are, and I make a decision (I don't want it to bite me), quickly calculate how close I am to the snake and its striking distance, and answer a question: Do I need to change my current direction, and speed? Yes, I should move back. A command is sent to put the muscles into gear and then do it. All this processing takes a long time, up to a second or two, and I could have been bitten while I was still in the midst of it. Luckily, however, all that doesn't have to occur. The brain takes a nonconscious shortcut through the amygdala, which sits under the thalamus and keeps track of everything streaming through. If a pattern associated with danger in the past is recognized by the amygdala, it sends an impulse along a direct connection to the brain stem, which then activates the flight-or-fight response and rings the alarm. I automatically jump back before I realize why. I did not make a conscious decision to jump, it happened without my conscious consent. This is more apparent after I have jumped back on my brother's foot, and my consciousness finally kicks in that it was not a snake, just the wind. This well-studied, faster pathway, the old fight-or-flight response, is present, of course, in other mammals, and has been honed by evolution.

If you were to have asked me why I had jumped, I would have replied that I thought I'd seen a snake. That answer certainly makes sense, but the truth is I jumped before I was conscious of the snake: I had seen it, but I didn't know I had seen it. My explanation is from post hoc information I have in my conscious system: The facts are that I jumped and that I saw a snake. The reality, however, is that I jumped way before (in the world of milliseconds) I was conscious of the snake. I did not make a conscious decision to jump and then consciously execute it. When I answered that question, I was, in a sense, confabulating: giving a fictitious account of a past event, believing it to be true. The real reason I jumped was an automatic nonconscious reaction to the fear response set into play by the amygdala. The reason I would have confabulated is that our human brains are driven to infer causality. They are driven to explain events that make sense out of the scattered facts. The facts that my conscious brain had to work with were that I saw a snake, and I jumped. It did not register that I jumped before I was consciously aware of the snake.

We are going to learn something strange about ourselves in this chapter. When we set out to explain our actions, they are all post hoc explanations using post hoc observations with no access to nonconscious processing. Not only that, our left brain fudges things a bit to fit into a makes-sense story. It is only when the stories stray too far from the facts that the right brain pulls the reins in. These explanations are all based on what makes it into our consciousness, but the reality is the actions and the feelings happen before we are consciously aware of them—and most of them are the results of nonconscious processes, which will never make it into the explanations. The reality is, listening to people's explanations of their actions is interesting—and in the case of politicians, entertaining—but often a waste of time.

THE UNCONSCIOUS ICEBERG

Consciousness takes time, which we don't always have. Our ancestors were those who were fast in life-threatening and competitive situations; the slow ones weren't around long enough to reproduce and didn't become ancestors. It is easy to show the difference in timing between automatic responses and those where consciousness intervenes. If I put you in front of a screen and have you push a button when a light flashes on, after a few trials you will be able to do this in about 220 milliseconds. If I ask you to slow this down just a tad, say to 240 or 250 milliseconds, you wouldn't be able to do it. Your speed would be more than 50 percent slower, it would drop to about 550 milliseconds. Once you put consciousness in the loop, your conscious self-monitoring of the speed takes longer, because consciousness works at a slower base speed. This is something that you may already be familiar with. Remember practicing the piano, or any other instrument, and memorizing a piece? Once you had practiced a piece, your fingers could really fly until you made a mistake and consciously tried to correct what you did wrong. Then, you could barely even remember what note was next. You were better off starting all over and hoping that your fingers would make it past the rough patch on their own. This is why good teachers warn their students not to stop when they make a mistake while playing in a recital, just keep on going, keep that automatic playing automatic. The same is true in sports. Don't think about that free throw, just plop it in as you have the hundreds of times in practice! "Choking" happens when consciousness steps into the play and throws the timing off.

Natural selection pushes for nonconscious processes. Fast and automatic is the ticket for success. Conscious processes are expensive: They require not only a lot of time, but also a lot of memory. Unconscious processes, on the other hand, are fast and rule-driven. Blatant examples of nonconscious processing can be readily seen in optical illusions. Our visual system takes in certain cues and automatically adjusts our perceptions to them. If you look at the shapes of the two tables presented in the illusion below, called the "turning tables illusion" produced by Roger Shepard, they are exactly the same area and shape. No one believes it! In fact, if you put this illusion into an introductory psychology textbook, students will cut out the pictures to see if they truly do match. Your brain is computing and adding corrections, adjusting to the visual cues of the orientation of the tables—and you cannot stop it. Even after you have cut and pasted the tables on top of one another and see that they are exactly the same size, you cannot consciously change the visual image to make them appear equal. Thus, when certain stimuli trick your visual system into constructing an illusion, even when you know you have been tricked, the illusion does not disappear. The part of the visual system that produces the illusion is impervious to correction based on conscious knowledge.[*]

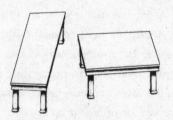

These tables look different, but actually they are the same exact size and shape. If you measure both, you will find them identical.

---

[*] To view this and other illusions, go to: http://michaelbach.de/ot/index.html

Some convincing illusions, however, can leave behavior unaffected. For example, while viewing the famous Müller-Lyer illusion, observers are asked to demonstrate with their fingers the size of a line presented with an arrowhead attached to each end, either both pointing in or out. Although the arrowheads can alter the perceived size of a line and deceive the eye (observers will typically say that the line with the outward pointing arrows is longer), observers do not make a corresponding adjustment in the distance between their fingers. It does not deceive the hand. This suggests that the processes determining the overt behavior are isolated from those underlying the perception. Thus, a visuomotor process responding to a visual stimulus can proceed independently of the simultaneous perception of that stimulus.[1] Things change, however, when you throw consciousness into the loop. Observers, who aren't asked to do the finger scaling until after a bit of time has elapsed, *do* make the adjustment.

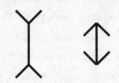

The Müller-Lyer illusion

Stimuli that are not consciously perceived, however, can affect behavior. For example, in one study, Stanislas Dehaene[2] and his colleagues in France, briefly (43 milliseconds) flashed either a prime (a stimulus that influences the subsequent response), which was either a number or the number written as a word to their volunteers, followed by a masking stimulus (two nonsense letter strings). The volunteers could not reliably report the prime's presence or absence, nor discriminate it from the nonsense strings. In other words, the primed number or word had not made it into their conscious awareness. Then the volunteers were flashed a target number and were told to press a response with one hand if the number was greater than five, the other hand if it were less. If the prime and the target number were both less than five (congruent), they responded faster than if they were incongruent. With brain imaging, these researchers showed that the prime, which never reached conscious awareness and went unperceived by the subject, actually activated the motor cortex. When you add to this the observation that robust perceptual after-effects can be induced by stimuli that are not consciously perceived,[3] it becomes evident that a great deal of the brain's work occurs outside of conscious awareness and control. (My brain made me do it!) Thus, the systems built into our brains carry out their jobs automatically when presented with stimuli within their domain, often without our conscious knowledge.

Automaticity can also be acquired. It comes with practice. In addition to the playing of musical instruments, another example is typing. After you are well practiced, you can type without even thinking about it. (And we have all read a few of those books!) If I ask you where the *V* is, however, you have to stop and consciously think about it. That is slow. Automating makes us so much more efficient. Automating processes is what makes us experts. Radiologists who read mammograms get more accurate and faster, the more mammograms they read. It is because the pattern recognition system in their brain has been trained and automatically recognizes the patterns of abnormal tissue. People become experts by developing automatic pattern recognition for a particular job.

WHY DO WE FEEL UNIFIED?

Now that we are aware (conscious!) that most of our processing is going on unconsciously and automatically, we arrive back at the question posed at the end of the last chapter. With so many complex systems going on subconsciously in a diversified and distributed way, why do we feel unified? I believe the answer to this question resides in the left hemisphere and one of its modules that we happened upon during our years of research. Once again our split-brain patients revealed some startling findings.

More than a few years into our experiments, we were working with another group of split-brain patients on the East Coast and began to wonder what these patients felt like when we would sneak information into their right hemisphere and tell the left hand to do something. What do they say to themselves when all of a sudden their left hand does something? It would be as if while you were reading this book, all of a sudden you saw your hand start snapping its fingers. How would you explain that to yourself? We set up an experiment where we were able to ask the patients what they thought their left hand was doing. These experiments revealed another capacity of the left hemisphere that stunned us.

We showed a split-brain patient two pictures: A chicken claw was shown to his right visual field, so the left hemisphere only saw the claw picture, and a snow scene was shown to the left visual field, so the right hemisphere only saw that. He was then asked to choose a picture from an array of pictures placed in full view in front of him, which both hemispheres could see. The left hand pointed to a shovel (which was the most appropriate answer for the snow scene) and the right hand pointed to a chicken (the most appropriate answer for the chicken claw). Then we asked why he chose those items. His left-hemisphere speech center replied, "Oh, that's simple. The chicken claw goes with the chicken," easily explaining what it knew. It had seen the chicken claw. Then, looking down at his left hand pointing to the shovel, without missing a beat, he said, "And you need a shovel to clean out the chicken shed." Immediately, the left brain, observing the left hand's response without the knowledge of why it had picked that item, put it into a context that would explain it. It interpreted the response in a context consistent with what it knew, and all it knew was: chicken claw. It knew nothing about the snow scene, but it had to explain the shovel in his left hand. Well, chickens do make a mess, and you have to clean it up. Ah, that's it! Makes sense. What was interesting was that the left hemisphere did not say, "I don't know," which truly was the correct answer. It made up a post hoc answer that fit the situation. It confabulated, taking cues from what it knew and putting them together in an answer that made sense. We called this left-hemisphere process *the interpreter*.[4]

We have numerous examples of this process at work in our split-brain patients. For instance, we flashed the words *bell* to the right brain and *music* to the left brain. The patient reported that he had seen the word *music*. When asked to point to a picture of what he just saw, our patient chose the bell, even though there were other pictures that better depicted music. Then we asked him: "Why did you pick the bell?" He replied, "Well, music, the last time I heard any music was the bells banging outside here." (He was referring to the bell tower.) His speaking left brain had to concoct a story to explain why he had pointed to the bell. In another experiment, we flashed the words *red* to the left hemisphere, and *banana* to the right hemisphere. Then we placed an assortment of different colored pens on the table and asked him to draw a picture with his left hand. He picked up the red pen (which was the left hemisphere making an easy decision), and he drew a banana with the left hand, (which was the right hemisphere). When I asked why he drew a banana, his left hemisphere, which had no clue why his left hand had drawn a banana, replied, "It is the easiest to draw with this hand because this hand can pull down easier." Once again, he did not say, "I don't know," which would have been the accurate answer.

We wondered if explanations of emotional responses or changes were also subject to this post hoc confabulation and gave the same type of test to a young teenage patient after inducing a mood shift. First we asked out loud: "Who is your favorite . . ." (both hemispheres heard that much) and then we lateralized to the right hemisphere only the word *girlfriend*. He immediately smiled, blushed, acted embarrassed (the mood shift), and shook his head, but stated that he didn't hear the word. He wouldn't say anything more. He had the right emotional response for a teenager being asked about a girlfriend, including reticence about discussing it, but he didn't know why. Eventually, he spelled out his girlfriend's name with his left hand.

DON'T PLAY AGAINST A RAT IN VEGAS!

We then wondered if we could create an experiment in which we could show that the left hemisphere and right hemisphere are different in the way each analyzes the world. We used a classic experimental psychology game, called a probability guessing experiment, where subjects guess which of two events will occur next: will the light flash above or below the line? The experimenter can manipulate the light so that it flashes above the line 80 percent of the time and below the line 20 percent of the time. It turns out that rats are better at this game than people are. Animals other than humans tend to maximize, that is, they always choose the option that has occurred the most frequently in the past. Rats quickly figure out to always guess above. That way they get a reward 80 percent of the time. Pigeons maximize. The house in Vegas maximizes. Kids under the age of four maximize.[5] But then something happens: Humans over the age of four use a different strategy, called frequency matching, where they tend to match the *frequency* of previous occurrences *in their guesses*. They guess above the line 80 percent of the time and below the line 20 percent of the time. The problem with that strategy is that since the order of occurrence is entirely random, it can result in a great deal of error. Even when told that the pattern is random, people try to figure out a system. On average, in the above situation, they only get the answer correct about 67 percent of the time. We devised a way to present this game to the two hemispheres separately and found that the right hemisphere is a maximizer,[6] just like the rats and pigeons and the four-year-old humans; it is the left hemisphere that is a frequency matcher. It tries to figure out a system; it is driven to infer a cause for the frequency of the flashes and creates a theory to explain them. We have concluded that the neural processes responsible for searching for patterns in events are housed in the left hemisphere. It is the left hemisphere that engages in the human tendency to find order in chaos, that tries to fit everything into a story and put it into a context. It seems that it is driven to hypothesize about the structure of the world even in the face of evidence that no pattern exists. It persists in this endeavor even when it is sometimes detrimental to performance—with slot machines, for instance.

It seems odd that the left hemisphere does this even when it can be nonadaptive. Why do we have such a system that can have such a deleterious effect on accuracy? Well, the answer is that for the most part it *is* adaptive, or we wouldn't have it. Patterns in the outside world often have discernable, deterministic causes, and having a system that seeks them has given us an edge everywhere but Vegas.

ON THE JOB WITH THE INTERPRETER

Once we understand that the left-brain interpreter process is driven to seek explanations or causes for events, we can see it at work in all sorts of situations. In fact, it can explain the observations of many past experiments. For instance, the results of a famous social psychology study, done in 1980, can be understood in light of the later discovery of this interpreter mechanism. In that experiment, the subjects had a prominent scar applied to their face with makeup, which they observed in a mirror.[7] They were told that they were going to have a discussion with another person, and that the experimenter was interested in whether the other person's behavior would be affected by the subject's disability, the scar. The subject was instructed to note any behavior that they thought was a reaction to the scar. At the last moment, the experimenter said he had to moisturize the scar to prevent it from cracking. What he really did, without the knowledge of the subject, was remove it. The subjects then had the discussion with the other person, and after it was over they were asked by the experimenter how it went. The subjects reported that they were treated horribly and that the other person was tense and patronizing. They were then shown a video of the other person taken during the discussion and asked to identify when the other person was reacting to the scar. As soon as the video started up, they'd stop it and point out that the other person looked away, attributing this to the scar, and so it went throughout the video. Their interpreter module grasped the first and easiest explanation it could make with the information that was available: there was a disfiguring scar, the other person frequently glanced away, there was no one else in the room, and there were no other distractions. Its makes-sense explanation was that the person looked away because of the scar. The interpreter was driven to infer cause and effect. It continually explains the world using the inputs that it has from the current cognitive state and cues from the surroundings. Interestingly, people normally glance away during conversations, but it usually goes unnoticed. This information that their conversation partner frequently glanced away only made it into the consciousness of these subjects because they were on guard for reactions and primed to notice them. Their whole story, the absolute reality for them at that moment, was based on two faulty pieces of information: (1) they had a scar, and (2) their conversation partner was glancing away more often than usual. So we must keep in mind that the interpreter's explanations are only as good as the information that it receives.

We use our interpreter module all day long, grasping the gist of situations, interpreting inputs and our body's physiological reactions, explaining all. In the last chapter we talked about how the right hemisphere lived a literal life and remembered exact items from a study set, whereas the left hemisphere tends to falsely recognize items that are similar as being the same. As I said earlier, it fudges. Our interpreter does this not only with objects, but with events. In one experiment, on healthy subjects who had not had split-brain surgery, we showed a series of about forty pictures that told a story of a man waking up in the morning, putting on his clothes, eating breakfast, and going to work. Then, after a bit, we tested each viewer on what pictures he had seen. This time he was presented with another series of pictures: Some of them were the original pictures, interspersed with some that weren't originally presented but could easily fit the story, and some distracter pictures that had nothing to do with the story, such as the man out playing golf or at the zoo. What you and I do, in such a task, is incorporate both the actual pictures and the related pictures, and we easily sort out the distracter pictures. In split-brain patients, this is also how the left hemisphere responds. The right hemisphere, however, does not do this. Just as we learned in the last chapter in remembering objects, it is totally veridical and only identifies the original pictures. The left brain gets the gist of story and accepts anything that fits in, but tosses out anything that does not. This elaboration has a deleterious effect on accuracy but usually makes it easier to process new information. The right brain does not infer the gist of the story; it is very literal and doesn't include anything that wasn't there originally. And this is why your three-year-old, embarrassingly, will contradict you as you embellish a story. The child is still maximizing, and the left-hemisphere interpreter, which is satisfied with the gist, is not fully in gear.

As I said, the interpreter is an extremely busy system. We found that it is even active in the emotional sphere, trying to explain mood shifts. In one of our patients, we triggered a negative mood in the right hemisphere by showing a scary fire safety video about a guy getting pushed into a fire. When asked what she saw, she said, "I don't really know what I saw. I think just a white flash." But when asked if it made her feel any emotion, she said, "I don't really know why, but I'm kind of scared. I feel jumpy, I think maybe I don't like this room, or maybe it's you, you're getting me nervous." She then turned to one of the research assistants and said, "I know I like Dr. Gazzaniga, but right now I'm scared of him for some reason." She *felt* the emotional response to the video—all the autonomic results—but had no idea what caused them. The left-brain interpreter had to explain why she felt scared. The information it received from the environmental cues were that I was in the room asking questions and that nothing else was wrong. The first makes-sense explanation it arrived at was that I was scaring her. What we have found so fascinating is that facts are great but not necessary. The left brain uses what it has and ad-libs the rest. The first makes-sense explanation will do, so in this case, the experimenter did it! The left-brain interpreter creates order out of the chaos presented to it by all the other processes spewing out information. We tried again with another emotion and another patient. We flashed a picture of a pinup girl to her right hemisphere, and she snickered. Once again she said that she saw nothing, but when we asked her why she was laughing, she told us we had a funny machine. This is what our brain does all day long. It takes input from other areas of our brain and from the environment and synthesizes it into a story. It also takes input from the body, as illustrated in the following classic experiment.

The hormone epinephrine is excreted by the adrenal glands and activates the sympathetic nervous system, resulting in an increased heart rate, contracted blood vessels, and dilated airways, and by doing so, increases the supply of oxygen and glucose to the brain and muscles. It produces hand tremors, facial flushing, palpitations, and anxiety. Our bodies excrete it under all sorts of circumstances, from the flight-or-fight response mentioned above and other short-term stress reactions, whether triggered by danger (falling out of a raft in the midst of white water), excitement (those moments before your favorite performer steps out onstage), or irritation from loud noises, heat, or other environmental stressors, such as your boss. In 1962 Stanley Schachter and Jerry Singer at Columbia University did an experiment (which used deception in its design and most likely would not be allowed today) to prove that emotional states are determined by a combination of physiological arousal and cognitive factors.[8] Volunteers were told that they were getting a vitamin injection to see if it had any effect on the visual system, but what they really received was an injection of epinephrine. Some of the subjects were told that the vitamin injection would cause side effects such as palpations, tremors, and flushing, and some were told that there were no side effects. After the injection of epinephrine, the volunteers were put into contact with a confederate who behaved in either a euphoric or an angry manner. The subjects who were informed about the "side effects" of the injection attributed their symptoms, such as a racing heart, to the drug. The subjects who were not informed, however, attributed their autonomic arousal to the environment. Those who were with the euphoric confederate reported being elated, and those with the angry confederate reported being angry. Here were three different, reasonable explanations for the physical symptoms; however, only one was correct: the injection of epinephrine. Once again this finding illustrates the human tendency to generate explanations for events. When aroused, we are driven to explain why. If there is an obvious explanation we accept it, as did the group informed about the effects of epinephrine. When there is not an obvious explanation, we generate one.

So this left-brain interpretive process that we have takes all the input, puts it together in a makes-sense story, and out it comes. As we have seen, however, the left hemisphere's explanations are only as good as the information it receives. And in many of the above examples we have seen that the information it received was faulty.

YOU'RE ONLY AS GOOD AS YOUR INPUT

The discovery of this mechanism now makes you wonder how often it goes astray. We can easily think of examples where we may have misinterpreted interactions with others. It is not so easy, however, to identify when we may have misinterpreted our own emotional responses, and even more difficult when they are faulty. A variety of emotional states and psychological disturbances are initially produced by endogenous errors in cerebral metabolism, such as those known to be associated with panic attacks. Such a biologically driven event that results in a surge of epinephrine produces a different felt state, which in turn must be interpreted. Most individuals don't say to themselves, "Gosh, my increased heart rate and sweating must be due to a malfunction in my cerebral metabolism. I better get that checked out." Most people's interpretive system would take cues from their own unique past and present psychological history and the current environmental cues to come up with an explanation: "My heart is pounding and I am sweating, I must be scared, and what can be scaring me must be . . . [looks around and sees a dog] a dog! I am afraid of dogs!" If the endogenous events mend through medication or natural events, the interpretations given to the altered mood state remain. They have been stashed in memory. This is how phobias can originate.

Not only is the interpreter interpreting what we are feeling and the reasons for our behaviors, but it also is interpreting what is going on inside the brain. We captured this serendipitously. While testing one of our patients, VP, unexpectedly we found that she was able to make some inferences that other split-brain patients could not. For instance, in our other split-brain patients, if we were to show the words *head* to one hemisphere and *stone* to the other, the patient would draw a picture of a head and a stone, whereas you or I would draw a picture of a headstone. Well, VP also drew a headstone. What was that all about? Because there is a self-cueing going on all the time to compensate for lost brain processing, throughout the testing of our split-brain and other neurological patients we have had to keep on the lookout for where the integration of information is occurring: inside or outside the body. For instance, split-brain patients may move their head so that visual stimuli from both visual fields enter into both hemispheres, or they may say something out loud so that the right hemisphere may get audio input from the left hemisphere. Further testing revealed that VP could not match pictures of shapes, sizes, or colors of figures from one hemisphere to the other, so it was not a simple transfer of visual information. If, however, she saw the words *red square,* they did transfer and she was able to pick out the red square with the other hemisphere. It turned out that her surgery had inadvertently spared a few of her anterior callosal fibers, which could be seen on an MRI. Those particular fibers allowed printed words to be transferred across the hemispheres, so her left hemisphere also saw the words that her right hemisphere saw. Thus, her interpreter had the input of both the words, *head* and *stone,* and put them together into one story. On the other hand patient JW had a complete split. He has no internal transfer of information; any transfer of information is always outside the body, but the transfer is clever and quick and it seems like it is occurring inside the brain. We flashed the word *car* to his left hemisphere and *1928* to his right hemisphere, and then asked him to draw what he had seen. He is a good artist and is a car guy. With his left hand (which was only informed by his right hemisphere and had seen the word *1928*) he drew a 1928 car! Somehow both hemispheres cooperated in their motor output to draw a car, but the cuing and integration was happening on the paper outside his body. As his left hand drew, his left hemisphere saw what was being drawn and influenced the process, but this did not occur inside his brain, but as result of the other hemisphere's external actions.

So while we have this rather precocious interpreter always explaining behaviors, thoughts, and emotions that are pouring out of this parallel distributed system of ours, the question arises, Is there an interpreter in the right hemisphere, too? Of course, as frequently happens when studying the brain, surprising results surface that have to be explained. As I mentioned earlier, the right hemisphere is a maximizer. We found, however, that the right hemisphere does frequency-match when presented with stimuli for which it is specialized, such as the visual task of facial recognition. In this task, either the left or right hemisphere was directed to guess whether the face it would be shown would have facial hair (30 percent of the faces presented had facial hair). In this experiment, the left hemisphere, which is not a specialist, responds randomly.[9] This suggested to us that one hemisphere cedes control of a task to the other hemisphere if the other hemisphere specializes in that task.[10] This is cued by one hemisphere to the other simply by a faster speed of response.

Some of the right hemisphere's specializations involve visual processing. Paul Corballis, studying split-brain patients in our lab, proposed that the right hemisphere has a visual interpreter dedicated to resolving the ambiguities of representing a three-dimensional world on the basis of a two-dimensional image inherent in spatial vision. In his *Treatise on Physiological Optics*, published posthumously in 1909, Hermann von Helmholtz first suggested that in order to get our 3-D view of the world, visual perception occurs by unconsciously inferring information from the 2-D retinal image. He proposed a startling idea that perception was basically a cognitive process that included not only the information from the retina, but also the experiences and goals of the perceiver. Corballis emphasizes that profound intelligence is required to create an accurate representation of the world from the information provided by the retinal image and suggests a right-brain "interpreter" process accomplishes this.[11]

Figuring out why we are tricked by some visual illusions, finding that not every visual illusion is seen by both hemispheres, and understanding the role that each hemisphere plays in visual processing has been part of the unraveling of the mysteries of the visual system. Corballis and colleagues found that while both hemispheres are equally good at lower-level visual processing (the first stages of processing visual stimuli), such as perceiving illusory contours (an illusion where contours are perceived, even though there is no line, luminance, or color change),[12] the right hemisphere is better than the left at a variety of visual tasks that involve advanced processing. The right hemisphere is easily able to do tasks that involve discriminations that are spatial in nature, such as detecting whether two images are identical or mirror-reversed, detecting small differences in line orientation,[13] and mental rotation of objects,[14] while the left hemisphere is dismal at these assignments. The right is also superior at temporal-discrimination tasks such as judgments as to whether two objects appear on a screen for equal or unequal amounts of time.[15] It also turns out that the right hemisphere is exceptional at perceptual grouping. For example, if you show partially drawn figures to the right hemisphere, it can easily guess what they are, but the left hemisphere can't guess until the figure is nearly completely drawn. Another example is illusory line motion. This occurs when a line is shown on a visual display in its entirety, all at once, but to the observer it appears to propagate from one end. This illusion can be manipulated both at the lower level and the higher level of visual processing. If a dot appears at one end just before the line appears, then the line appears to propagate from the dot.[16] This is lower-level processing and the illusion is experienced by both hemispheres. If the line flashes between two dots of different colors or widths, it appears to propagate from the dot that it matches.[17] This involves higher-level processing; the right hemisphere sees this illusion, but the left does not.[18]

If the right brain is good at apprehending complex patterns and gets automatic about it, maybe, we thought, we could see it unfold in the abilities of grandmaster chess players. Chess players have often been the target of cognitive scientists, starting in the 1940s with the studies of Adriaan de Groot, a psychologist and chess player himself. International Grandmaster and two-time U.S. chess champion Patrick Wolff, who at age twenty defeated the world chess champion Gary Kasparov in twenty-five moves, came to our lab. We gave him five seconds to look at a picture of a chessboard with all the pieces set in a pattern that makes chess sense, and then asked him to reproduce it. He quickly and accurately did so, getting twenty-five out of twenty-seven players in the correct position. If you and I were to do this, we would, even if we were good players, only get about five pieces correctly placed. One question remained, however. Was he able to do this just because he has a very good visual memory? If that were true, then it shouldn't matter if the pieces were in positions that made chess sense or not. Back to the chessboard: He had another quick look at the same board, the same number of pieces, but in positions that didn't make chess sense. This time he only got a few pieces right, just like a non-chess-playing person. His original accuracy was from his right brain's automatically matching up patterns that it had learned from years of playing chess.

So while we, as neuroscientists, knew that Wolff's right-brain pattern perception mechanism is all coded, ran automatically, and was the source of this capacity, he did not. When he was asked about it, his left-brain interpreter struggled for an explanation. "[Y]ou sort of get it by trying to . . . to understand what's going on quickly, and of course you chunk things, right? I mean obviously, these pawns, just . . . but, but it, I mean, you chunk things in a normal way, like . . . I mean, one person might think this is sort of a structure, but actually I would think this is more, all the pawns like this. . . ."

The interpreter is only as good as the information it gets. The interpreter receives *the results* of the computations of a multitude of modules. It does not receive the information that there are multitudes of modules. It does not receive the information about how the modules work. It does not receive the information that there is a pattern-recognition system in the right hemisphere. The interpreter is a module that explains events from the information it does receive. So in Patrick Wolff's case, the information it received was that he can quickly replicate a chessboard at a glance, and that he has a considerable knowledge of chess. So that was what he used to explain his ability.

HIJACKING THE INTERPRETER

This concept that the interpreter is only as good as the data it receives is crucial in explaining many seemingly inexplicable behaviors of both normal brains and neurological patients. Indeed, if you feed the interpreter incorrect data you can hijack it. By doing this, a different story results than it may otherwise have produced. So perhaps, for our interpreter process, reality is virtual. It depends on the sensory cues that are here and now.

For instance, if you with your normally functioning brain were to go to a virtual reality lab, you would notice that the lab is a large room with a flat, concrete floor. That is your current reality. Then you put on the virtual reality glasses and what you see is controlled by the guy sitting over in the corner running the computer, who is happy to play tricks on you. You start to walk and all of a sudden a deep, gaping pit pops up in front of you. Yikes! You get a jolt of adrenaline, your heart races, and you jump back. You hear laughter. But just then a narrow plank appears across the pit and you are asked to walk across it. If you are like me, you will refuse, saying "NO WAY!" If you are a thrill seeker and you do try it, you will have your arms out for balance and will proceed at a snail's pass, heart thumping, and muscles tense. Of course, everyone else in the lab is laughing harder because you are on the flat, concrete floor. Even though you know this, however, your common sense has been hijacked by the perceptions of the moment. Your interpretation of the world is immediately influenced by the visual cues that have overridden what your conscious brain knows.

The interpreter is receiving data from the domains that monitor the visual system, the somatosensory system, the emotions, and cognitive representations. But as we just saw above, the interpreter is only as good as the information it receives. Lesions or malfunctions in any one of these domain-monitoring systems lead to an array of peculiar neurological conditions that involve the formation of either incomplete or delusional understandings about oneself, other individuals, objects, and the surrounding environment, manifesting in what appears to be bizarre behavior. It no longer seems bizarre, however, once you understand that such behaviors are the result of the interpreter getting no, or bad, information. In the last chapter we saw what happens when there was a lesion in the domain that monitors part of the visual system. Moving on to the domain that monitors the somatosensory system, a lesion there can produce a syndrome called anosognosia. A person with this syndrome will deny that their paralyzed left hand is theirs. Vilayanur Ramachandran recounts this encounter with such a patient:

PATIENT: [*pointing to her own left hand*] Doctor, whose hand
is this?

DOCTOR: Whose hand do you think it is?

PATIENT: Well, it certainly isn't yours!

DOCTOR: Then whose is it?

PATIENT: It isn't mine, either.

DOCTOR: Whose hand do you think it is?

PATIENT: It is my son's hand, Doctor.[19]

The parietal cortex is constantly seeking information on the arm's
position in three-dimensional space and also monitors the arm's existence in relation to everything else. If there is a lesion in the sensory
nerves in the periphery of the nervous system, the information flow to
the brain is interrupted. The monitoring system is not receiving information about where the arm is, what is in its hand, whether it is in
pain, feels hot or cold, or whether it can move or not. The monitoring
system puts up a squawk: "There is no input! Where's the left hand?"
But if the lesion is in the parietal cortex itself, then no monitoring goes
on and no squawk is raised because the squawker is damaged. A patient with a right parietal lesion suffers damage to the area that represents the body's left half. It is as if that part of the body has lost its
representation in the brain and left no trace, leaving no brain area to
report about the left half of the body and whether it is working or not
to the interpreter. For that patient, the left half of the body ceases to
exist. When a neurologist holds the patient's left hand up to her face,
no somatosensory information reaches the patient's interpreter and she
gives a reasonable response: "That's not my hand." The interpreter,
which is intact and working, is not receiving a report from the parietal
lobe about a left hand, therefore it cannot be hers. In this light, the
claims of the patient are more reasonable.

Another odd condition is Capgras' syndrome, where the problem is in
the system that monitors emotions. These patients will recognize a
closely related person but will insist that the person is an imposter and

has been replaced by an identical double. For instance, another patient described by Ramachandran, said about his father, "He looks exactly like my father, but he really isn't. He's a nice guy, but he isn't my father, Doctor." When asked why the man was pretending to be his father, he replied, "That is what is so surprising, Doctor—why should anyone want to pretend to be my father? Maybe my father employed him to take care of me—paid him some money so that he could pay my bills. . . ."[20] In this syndrome the emotional feelings for the familiar person are disconnected from the representation of that person.[21] The patient feels no emotion when they see the familiar person, and this can be measured by skin conductance responses. The interpreter has to explain this phenomenon. It is receiving the information from the face identification module: "That's Dad." However, it is not receiving any emotional information. It has to make a causal inference. The interpreter comes up with a solution: "It must not really be Dad, because if it really were Dad I'd feel some emotion, so he is an imposter!"

While such examples of a hijacked interpreter system may be fascinating, there are examples much closer to home. Drugs that reduce anxiety are commonly taken, but anxiety is not always a bad thing. If you are walking down the street and see someone acting suspiciously, it would be both normal and practical for you to feel some anxiety, arousal, and increased wariness. That burst of adrenaline has proven successful over hundreds of thousands of years of evolution. If you are taking a drug to suppress anxiety, however, you don't get that increased arousal and wariness when you see a menacing situation. Your monitoring system has been hijacked and feeds the interpreter bad information. You don't feel anxious and your interpretive system doesn't classify the situation as dangerous; it makes a different interpretation and you don't take special care. It has been claimed the increased use of such drugs in New York City is correlated with increased muggings and ER visits.

At other times an antianxiety drug is not the culprit that waylays our fight-or-flight response; rather, it is our interpreter rationalizing situations: "Calm down, nothing's strange, he is just a homeless guy." Perhaps having listened too often to advice about not being so suspicious of strangers and to be politically correct, the interpreter ignores warning signals.

Ramachandran suggested that various defense mechanisms, such as rationalization (creating fictitious evidence or false beliefs) and repression, arise because the brain arrives at the most probable and globally consistent interpretation of evidence derived from multiple sources, and then ignores or suppresses conflicting information. This suggestion is consistent with our finding that the left hemisphere frequency-matches and falsely identifies similar but novel stimuli as being the same as previously seen stimuli. It gets the gist of the situation from all the input, tries to find a pattern, and puts it together in a makes-sense interpretation. He also suggested that the right parietal lobe has a system that he calls an anomaly detector, which squawks when the discrepancies get too large. The literal right brain chimes in. This would account for the observation that patients with right parietal lobe lesions can have such outrageous, no-holds-barred stories coming out of their left hemisphere, unconstrained by their right-hemisphere anomaly detector, while this doesn't happen with left-hemisphere lesions, where the right, totally accurate, and exacting system is fully operative. Patients with left frontal lobe lesions tend not to be able to engage in denial, rationalization, or confabulatory "gap-filling" and often become depressed. Imagine if you could never come up with a rationalization to eat the chocolate cake.

I CAN'T BELIEVE MY EYES!

In moment-to-moment activity, the interpreter is always dealing with the changing input from sites in the brain where activity is going on. Sitting under that apple tree, Isaac Newton, indulging in that most human trait to constantly seek explanations and the causes for things, asked himself, "Why did the apple fall down? Hmmmm . . . nothing pushed it. Why doesn't it go up?" Newton was engaging in two differ-

ent types of processing concerned with causality, and we have found
that one type occurs in the right hemisphere, and the other in the left.
Albert Michotte, a Belgian experimental psychologist, came up with
the best known example of perceptual causality, known as Michotte's
balls. If after observing a green ball on a screen move toward a red ball,
stop when it contacts it, and then the red ball immediately moves away,
most people report that the green ball caused the red ball to move.
This is perceptual causality: It is the direct perception, in this case by
observation, that some action occurred as a result of physical contact.
If, however, a time gap takes place between when the balls contact
each other and when the red ball moves off, or if the balls don't actually
touch and the red ball moves off, most report that there is no causal
relationship. It is the right hemisphere that can see this difference.[22]
The left hemisphere is unaffected by the time or space gap when it
reported causality and reports that the green ball caused the red one to
move in all three cases. Perceptual causality is the bailiwick of the
right hemisphere. So when Newton observed that the apple fell but
perceived no observable interaction that caused it, he was using his
right hemisphere. For other animals, that is the end of the story. But it
wasn't good enough for Newton. He went on to employ causal inference, the application of logical rules and conceptual knowledge to the
interpretation of events, which, as you may have guessed, is the bailiwick of the left hemisphere. This can be seen in the results of the following experiment: Two small boxes, a red one and a green one, were
suspended above a larger box. When one or the other of these boxes
dropped down and touched the larger box, either independently or together, the larger box would light up only if it were touched by the
green box. The left hemisphere can make the causal inference right
away that the larger box must be touched by the green box for it to light
up, but the right hemisphere simply can't do it. As you and I muddle
through life, moving from one task to the next, different regions, distributed across the brain, come into play and are seamlessly blended
together, dominating our consciousness from moment to moment.

ON THE FLY WITH THE INTERPRETER

We saw this blending in action when, unexpectedly, first one, and later a few others, of our split-brain patients began to speak a few words from their right brain. We were stunned when we flashed two words, *key,* to the right hemisphere, and *fork,* to the left, and the patient, instead of just saying "fork," said "fork" and then said "key." Now what was going on? So once again we wanted to know if there was transfer of information between the hemispheres, either internally or externally, or if both hemispheres were talking. To suss this out, we flashed two more pictures, but this time instructed the patient not to tell us what they were, but if they were the same or different. She could not do this. After more testing it was apparent that the right hemisphere was tossing out a word and there was no transfer of information. We began to do some tests that showed how quickly one adapts, the interpreter just grabbing whatever information it can. We showed the patient PS a series of five slides with two words on each slide (Mary+ Ann; May + Come; Visit + Into; The + Town; Ship+ Today). The word on the left was seen by the right hemisphere and the word on the right was seen by the left hemisphere. The five words that each hemisphere saw made sense as a story. The right hemisphere saw: Mary may visit the ship. The left hemisphere saw: Ann come into town today. Read normally from left to right, as you and I would do, the series of slides makes a story: Mary Ann may come visit into the township today. So what does the split-brain patient say he saw?

> PS: Ann come into town today [*the left hemisphere answers*]
> EXPERIMENTER (E): Anything else?
> PS: On a ship [*here comes the right hemisphere*]
> E: Who?
> PS: Ma
> E: What else?

PS: To visit

E: What else?

PS: To see Mary Ann

E: Now repeat the whole story

PS: Ma ought to come into town today to visit Mary Ann
on the boat.[23]

PS was interweaving the words after he spoke them. The interpreter received the information from the right hemisphere externally; it did not have access to this part of the story until it was uttered by the right hemisphere, heard by the left hemisphere, and then the interpreter had to deal with the situation. Once again we see the integration of disparate behaviors into a coherent framework. Order has been made from chaos. In doing so, behaviors originating from the right hemisphere were being incorporated into the conscious stream of the left hemisphere and we could see/hear it happening right in front of our eyes.

In another case we had flashed a picture of a Radio Flyer wagon to the right hemisphere. Out popped the word *toy* from the right hemisphere. In the following conversation, the left hemisphere, which didn't see the picture, has a hard time trying to explain why he said "toy":

E: Why does *toy* come to mind?

P: I don't know, the only thing that comes to mind. The first thing that bangs into my head.

E: Does it kind of look like a toy?

P: Yeah, that is what it feels like. It is almost like an inner sense tells you.

E: How often do you go with an inner sense and how often do you go with what things look like?

P: If I can't really tell what something looks like first thing, if I say what it is first thing, then I just go with that . . . the first thing that pops into my mind.

These rather blatant examples showed us that our cognitive system is not a unified network with a single purpose and single train of thought.

WHAT DOES IT ALL MEAN FOR THE BIG PICTURE?

The view in neuroscience today is that consciousness does not constitute a single, generalized process. It is becoming increasingly clear that consciousness involves a multitude of widely distributed specialized systems and disunited processes,[24] the products of which are integrated in a dynamic manner by the interpreter module. Consciousness is an emergent property. From moment to moment, different modules or systems compete for attention and the winner emerges as the neural system underlying that moment's conscious experience. Our conscious experience is assembled on the fly, as our brains respond to constantly changing inputs, calculate potential courses of action, and execute responses like a streetwise kid.

So, here we are, back to the leading question of the chapter: How come we have that powerful, almost self-evident feeling that we are unified when we are comprised of a gazillion modules? We do not experience a thousand chattering voices, but a unified experience. Consciousness flows easily and naturally from one moment to the next with a single, unified, and coherent narrative. The psychological unity we experience emerges out of the specialized system called "the interpreter" that generates explanations about our perceptions, memories, and actions and the relationships among them.[25] This leads to a personal narrative, the story that ties together all the disparate aspects of our conscious experience into a coherent whole: order from chaos. The interpreter module appears to be uniquely human and specialized to the left hemisphere. Its drive to generate hypotheses is the trigger for human beliefs, which, in turn, constrain our brain.

The constructive nature of our consciousness is not apparent to us.

The action of an interpretive system becomes observable only when the system can be tricked into making obvious errors by forcing it to work with an impoverished set of inputs, most obviously in the split-brain or in lesion patients, but also in normal patients who have been fed faulty information. Even in the damaged brain, however, this system still lets us feel like "us." We have learned from our split-brain patients that even when the left brain has lost all consciousness about the mental processes managed by the right brain and vice versa, the patient does not find one side of the brain missing the other. It is as if we don't have knowledge about what we no longer have access to. The emergent conscious state arises out of separate mental systems, and if they are disconnected or damaged there is no underlying circuitry from which the emergent property arises.

Our subjective awareness arises out of our dominant left hemisphere's unrelenting quest to explain these bits and pieces that have popped into consciousness. Notice that *popped* is in the past tense. This is a post hoc rationalization process. The interpreter that weaves our story only weaves what makes it into consciousness. Because consciousness is a slow process, whatever has made it to consciousness has already happened. It is a fait accompli. As we saw in my story at the beginning of the chapter, I had already jumped before I realized whether I had seen a snake or if it was the wind rustling the grass. What does it mean that we build our theories about ourselves after the fact? How much of the time are we confabulating, giving a fictitious account of a past event, believing it to be true?

This post hoc interpreting process has implications for and an impact on the big questions of free will and determinism, personal responsibility and our moral compass, which we will look at in the next chapter. When thinking about these big questions, one must always remember, *remember*, REMEMBER that all these modules are mental systems selected for over the course of evolution. The individuals who possessed them made choices that resulted in survival and reproduction. They became our ancestors.

Chapter Four

ABANDONING THE CONCEPT OF FREE WILL

THE HUMAN INTERPRETER HAS SET US UP FOR A FALL. IT has created the illusion of self and, with it, the sense we humans have agency and "freely" make decisions about our actions. In many ways it is a terrific and positive capacity for humans to possess. With increasing intelligence and with a capacity to see relationships beyond what is immediately and perceptually apparent, how long would it be before our species began to wonder what it all meant—what was the meaning of life? The interpreter provides the storyline and narrative, and we all believe we are agents acting of our own free will, making important choices. The illusion is so powerful that there is no amount of analysis that will change our sensation that we are all acting willfully and with purpose. The simple truth is that even the most strident determinists and fatalists at the personal psychological level do not actually believe they are pawns in the brain's chess game.

Puncturing this illusionary bubble of a single willing self is difficult to say the least. Just as we know but find it difficult to believe that the world is not flat, it too is difficult to believe that we are not totally free agents. We can begin to understand the illusion about free will when we ask the question, What on earth do humans want to be free from? Indeed, what does free will even mean? However actions are caused, we want them to be carried out with accuracy, consistency, and purpose. When we reach for the glass of water, we don't want our hand suddenly rubbing our eye, or grasping so hard that the glass shatters, or the water to spurt upward from the faucet or turning into mist. We want all the physical and chemical forces in the world to be on our side, serving our nervous and somatic systems so that whatever the job, it gets done right. So we don't want to be free from the physical laws of nature.

Think about the problem of free will on a social level. While we believe we are always acting freely, we commonly want none of that in others. We expect the taxi driver to take us to our destination and not where he thinks we ought to go. We want our elected politicians to vote on future issues the way we have decided (probably erroneously) they think. We don't like the idea they are freely wheelin' and dealin' when we send them off to Washington (though they probably are). We intensely desire reliability in our elected officials and indeed in our family and friends.

When all the great minds of the past dealt with the question of free will, the stark reality and clarity that we are big animals, albeit with unique attributes, was not fully appreciated and accepted. The powerful idea of determinism, however, was apparent and appreciated. At the same time, and prior to the startling advances in neuroscience, explanations of mechanisms were unknown. Today they are. Today we know we are evolved entities that work like a Swiss clock. Today, more than ever before, we need to know where we stand on the central question of whether not we are agents who are to be held accountable and responsible for our actions. It sure seems like we should be. Put simply: The issue isn't whether or not we are "free." The issue is that there is no scientific reason not to hold people accountable and responsible.

As we battle through this, I will attempt to make two main points: First—and this has to do with the very nature of brain-enabled conscious experience itself—we humans enjoy mental states that arise from our underlying neuronal, cell-to-cell interactions. Mental states do not exist without those interactions. At the same time, they cannot be defined or understood by knowing only the cellular interactions. Mental states that emerge from our neural actions do constrain the very brain activity that gave rise to them. Mental states such as beliefs, thoughts, and desires all arise from brain activity and in turn can and do influence our decisions to act one way or another. Ultimately, these interactions will only be understood with a new vocabulary that captures the fact that two different layers of stuff are interacting in such a way that existing alone animates neither. As John Doyle at Caltech puts the issue, "[T]he standard problem is illustrated with hardware and software; software depends on hardware to work, but is also in some sense more 'fundamental' in that it is what delivers function. So what causes what? Nothing is mysterious here, but using the language of 'cause' seems to muddle it. We should probably come up with new and appropriate language rather than try to get into some Aristotelian categories." Understanding this nexus and finding the right language to describe it represents, as Doyle says, "the hardest and most unique problem in science."[1] The freedom that is represented in a choice not to eat the jelly donut comes from a mental layer belief about health and weight, and it can trump the pull to eat the donut because of its yummy taste. The bottom-up pull sometimes loses out to a top-down belief in the battle to initiate an action. And yet the top layer does not function alone or without the participation of the bottom layer.

The second point is how to think about the very concept of personal responsibility in a mechanistic and social world. It is a given that all network systems, social or mechanical, need accountability in order to work. In human societies this is generally referred to as members of a social group possessing personal responsibility. Now is personal responsibility a mechanism that resides in the individual brain? Or is its existence dependent on the presence of a social group? Alternatively, does the concept have meaning only when considering actions within a social group? If there were only one person in the world, would the concept of personal responsibility have any meaning? I would suggest it would not and in that truth, one can see that the concept is wholly dependent on social interactions, the rules of social engagement. It is not something to be found in the brain. Of course, some concepts that would lack meaning if nobody else were around are not wholly dependent on social rules or interactions. If there were only one person, it would be meaningless to say that he is the tallest person or taller than everyone else, but the concept of "taller" is not wholly dependent on social rules.

One cannot emphasize enough how all of this seems like crazy academic intellectual talk. It seems like when I go to a restaurant, my meal selection is a free choice. Or when the alarm goes off in the morning, I can go exercise or roll over, but it is my free choice. Or on the other hand, I can walk into a store and choose not to slip something into my pocket without paying for it. In traditional philosophy, free will is the belief that human behavior is an expression of personal choice that is not determined by physical forces, Fate, or God. YOU are calling the shots. YOU, a self with a central command center, are in charge, are free from causation, and are doing things. You can be free from outside control, coercion, compulsion, delusion, and inner lack of restraint over your actions. From what we learned in the last chapter, however, the modern perspective is that brains enable minds, and that YOU is your vastly parallel and distributed brain without a central command center. There is no ghost in the machine, no secret stuff that is YOU. That YOU that you are so proud of is a story woven together by your interpreter module to account for as much of your behavior as it can incorporate, and it denies or rationalizes the rest.

We have seen that our functionality is automatic: We putter along perceiving, breathing, making blood cells, and digesting without so much as a thought about it. We also automatically behave in certain ways: We form coalitions, share our food with our children, and pull away from pain. We humans also automatically believe certain things:

We believe incest is wrong and flowers aren't scary. Our left-brain interpreter's narrative capability is one of the automatic processes, and it gives rise to the illusion of unity or purpose, which is a post hoc phenomenon. Does this mean we are just along for the ride, cruising on autopilot? Our whole life and everything that we do or think is determined? Oh my. As I already said, with what we now know about how the brain operates, it seems that we need to reframe the question about what it means to have free will. What on earth are we really talking about anyway?

NEWTON'S UNIVERSAL LAWS AND MY HOUSE

In 1975, perhaps not deliberating long enough on my decision, I chose to build my house and I did. Notice that I did not say I chose to have my house built, which, perhaps, would have yielded a better result. For years I was the brunt of jokes concerning the fact that a ball placed on the floor of the living room would roll, unaided, across the dining room and into the kitchen. Similar phenomena were observed on the kitchen counter. Those who were bothered by lines that were not straight would also comment upon the windows across the front of the house. My house was a house that a physicist should have loved, for not only did it readily illustrate Newton's laws of motion and some principles of chaos theory, but they could also point to it and laugh that obviously it was built by someone from the biological side of science, someone who was comfortable with inexact measurements, obviously not an engineer.

First of all, my house demonstrated a basic principle of experimental science: No real measurement is infinitely precise; it always includes a degree of uncertainty in the value—wiggle room. Uncertainty is present because no matter what measuring device is used, it has a finite precision and, therefore, imprecision, which can never be eliminated completely, even as a theoretical idea. In fact, in some cases the actual action of measuring something can change its measurement. Physicists know this but don't like it. That is why they keep inventing more and more precise measuring apparatuses, and I should have used more of them. I admit it, in building my house there were some imprecise measurements initially. While physicists will nod here that, yes, deplorable as it is, it is to be expected, my son-in-law, who is a contractor, would be rolling his eyes. And so would Isaac Newton, because thanks to that seventeenth-century scientist, physicists for some two centuries thought it would be possible to finally get the perfect measurement, and, once you had it, everything should fall neatly into place. Plug in a number at the beginning of an equation, and you will always get the same answer at the end.

Newton was no slacker as a student. While he was attending Cambridge University, it was hit by the plague and closed for two years. Instead of sitting by the fire reading novels (maybe Chaucer), playing billiards, and drinking beer to while away the time until school opened again, he read Galileo and Kepler and invented calculus. This turned out to be a good thing, because it came in handy a few years later. The Italian astronomer Galileo Galilei, who had died in 1643, the year Newton was born, was the original "just do it" guy. Instead of sitting around talking about how he thought the universe was constructed (Plato's modus operandi), he decided to back up his ideas and observations with measurements and mathematics. It was Galileo who came up with the big ideas that objects retain their velocity and straight-line trajectories unless a force (often friction) acts upon them, as opposed to Aristotle's hypothesis that objects naturally slow and stop unless a force acts upon them to keep them going. He also came up with the idea of inertia (the natural resistance of an object to changes in motion) and identified friction as a force.

Newton put these ideas all together in one tidy package. After scrutinizing the experimental observations and data of Galileo, Newton wrote down Galileo's laws of motion as algebraic equations and realized these equations also described Kepler's observations about planetary motion. This had not dawned on Galileo. Newton came up with the notion that the physical matter of the *universe*—that would be everything—operated according to a set of fixed, knowable laws, mathematical relationships that he had just jotted down. His three laws of motion, which governed the balls in my living room, have stood the test of more than three centuries of experimentation and practical application, from clocks to skyscrapers. Newton, however, rocked the world with his laws, not just the hallowed halls of physics. Why, you may wonder, did some guy messing around with calculus, Galileo's data, and apples create such a stir? If you were like me, physics class didn't really put you into any existential crisis.

Determinism

If the topic of determinism were to be brought up at dinner, the finger would most likely be pointed at Newton and his universal laws, although the idea had been floating around since the time of those inquisitive Greeks. Newton had reduced the machinations of the universe into a set of mathematical formulas. If the universe's machinations followed a set of determined laws, then, well, everything is determined at the get-go. As I said earlier, determinism is the philosophical belief that all current and future events, actions, including human cognition, decisions, and behavior are *causally* necessitated by preceding events combined with the laws of nature. The corollary, then, is that every event, action, et cetera, is predetermined and can in principle be predicted in advance, if all parameters are known. Newton's laws also work in reverse. That means that time does not have a direction. So you can also know something's past by looking at its present state. (As if the free will and determinism issue were not numbing enough, some serious philosophers and physicists believe that time itself does not exist. The argument is that it too is an illusion. All of this plays out on the phenomenological backdrop that humans feel free in real time.) Determinists believe that the universe, and everything in it, is completely governed by causal laws. Have their left-hemisphere interpreters run amok and made it to prime time? After we get a little more physics under our belt, we will come back to this idea of causation.

Now the ramifications of this idea are disturbing to just about everyone. If the universe and everything in it are following predetermined laws, then that seems to imply that individuals are not personally responsible for their actions. Go ahead and eat the Death by Chocolate cake, it was preordained about two billion years ago. Cheat on the test? You have no control over that—go ahead. Not getting along with your husband? Slip him some poison and say the universe made you do it. This is what caused such a stir when Newton presented his universal laws. I call this the Bleak View, but many scientists and determinists think this is the way things are. The rest of us just don't believe it. "The universe made me buy that dress!" or "The universe made me buy that Boxster!"* just isn't going to fly well at the dinner table. If we were to be logical neuroscientists, however, shouldn't it?

A Post Hoc World?

We accept the idea that our bodies are humming along, being run by automatic systems that follow deterministic laws. Luckily, we don't have to consciously digest our food, keep our heart beating, and our lungs oxygenating. When it comes to our thoughts and actions, however, we don't like to think of those as being nonconscious, following a set of predetermined laws. But the fact remains, and you can show this experimentally, that actions are over, done, kaput, before your brain is conscious of them. Your left-hemisphere interpretive system is what pushes the advent of consciousness back in time to account for the cause of the action. The interpreter is always asking and answering the question, WHY? In fact, Hakwan Lau, now at Columbia University, can mess with this misconception of timing in your brain. He

---

* Thanks to Flip Wilson's famous album *The Devil Made Me Buy That Dress!*

was looking to see whether he could prove or disprove whether conscious control of actions was illusory or real by using transcranial magnetic stimulation (TMS).

TMS does what the name implies. Plastic-enclosed coils of wire are placed on the outside of the head. When activated, a magnetic field is produced that passes through the skull and induces a current in the brain that locally activates the nerve cells. This can be applied to specific cells or to an area generally and thus the functions and connections of different parts of the brain can be studied. Activity of parts of the brain can also be inhibited, and so one may study what a specific area does when it is disconnected from the processes of other areas. The area of the frontal cortex called the supplemental motor area (SMA) is involved with the planning of motor actions that are sequences of action done from memory, such as playing a memorized piano prelude. The pre-SMA is the area that is involved with acquiring new sequences. Lau knew, from the work of others, that stimulation of the medial frontal cortex gives one the feeling of the urge to move[2] and that lesions in this area in macaque monkeys abolish their self-initiated movements.[3] He, himself, had previously found that there is activation in this area when subjects generated actions of their own free choice.[4] The pre-SMA was, thus, his area of interest. Lau found that when TMS is applied over the pre-SMA *after* the execution of a spontaneous action, the *perceived* onset of the *intention* to act, that moment when you become conscious that you intend to act, is shifted *backward* in time on the temporal map,* and the *perceived* time of the actual *action,* the moment when you are conscious that you are acting, is shifted *forward* in time.[5] What I think he has done is actually mess with the interpreter module.

While the idea that there is a temporal map that intentions and actions are mapped onto, but not necessarily as they actually happened,

---

* Brain maps are neuronal representations of the world, one of which is for time, or the temporal map.

seems crazy, it happens to you all the time. Think about when you smash your finger with a hammer and pull it away. Your explanation will be that you smashed your finger, it hurt, and you pulled it away. Actually, however, what happens is you pull it away before you feel the pain. It takes a few seconds for you to perceive, or be conscious of, the pain and your finger has long since gotten out of Dodge. What has happened is the pain receptors in your finger send a signal along the nerve to the spinal cord, and immediately a signal is sent back along motor nerves to your finger, triggering the muscles to contract and pull away without involving the brain, a reflexive action. You move first. The pain receptor signal is also sent up to the brain. Only after the brain processes the signal and interprets it as pain do you become conscious of the pain. Consciousness takes time, and it was not consciousness of the pain followed by a conscious decision that moved your finger: Pulling your finger back was a reflex and was done automatically. The signal that produces awareness of pain originates in your brain after the injury and is referred to the finger, but your finger has already moved. Your interpreter has to put all the observable facts together (pain and moved finger) in a makes-sense story to answer the WHY? question. It makes sense that you pulled your finger away because of the pain, so it just fudges the timing. In short, the interpreter makes the story fit with the pleasing idea one actually willed the action.

The belief that we have free will permeates our culture, and this belief is reinforced by the fact that people and societies behave better when they believe that is the way things work. Is a belief, a mental state, constraining the brain? Kathleen Vohs, a psychology professor at the Carlson School of Management in Minnesota, and Jonathan Schooler,[6] a psychology professor at the University of California–Santa Barbara, have shown in a clever experiment that people act better when they believe they have free will. Curious that in a huge survey of people in 36 countries, more than 70 percent agreed that their life was in their own hands, and also knowing that other studies have shown that changing people's sense of responsibility can change their behavior,[7] Vohs and Schooler set about to see empirically whether people work better when they believe that they are free to function. College students were given a passage from Francis Crick's book *The Astonishing Hypothesis* which has a deterministic bias, to read before taking a computerized test. They were told that there was a glitch in the software, and that the answer to each question would pop up automatically. They were instructed that, to prevent this from happening, they had to push one of the computer keys and were asked to do so. Thus it took extra effort not to cheat. Another group of students read an uplifting book with a positive outlook on life, and they also took the test. What happened? The students who read about determinism cheated, while those who had read the positive attitude book did not. In essence, one mental state affected another mental state. Vohs and Schooler suggested that disbelief in free will produces a subtle cue that exerting effort is futile, thus granting permission not to bother.

People prefer not to bother, because bothering, in the form of self-control, requires exertion and depletes energy.[8] Further investigation along these lines by Florida State University social psychologists Roy Baumeister, E. J. Masicampo, and C. Nathan DeWall found that reading deterministic passages increased tendencies of the people they studied to act aggressively and to be less helpful toward others.[9] They suggest that a belief in free will may be crucial for motivating people to control their automatic impulses to act selfishly, and a significant amount of self-control and mental energy is required to override selfish impulses and to restrain aggressive impulses. The mental state supporting the idea of voluntary actions had an effect on the subsequent action decision. It seems that not only do we believe we control our actions, but it is good for everyone to believe it.

At the level of university life, however, there has been an assault for the last several centuries on the idea of free will from the determinists. Stirring things up in the sixteenth century, Copernicus declared that the Earth was not the center of the universe, followed up as we know by Galileo and Newton. Later, René Descartes, although more famous for a dualist stance, proposed that the bodily functions followed bio-logical rules; Charles Darwin put forth his evolutionary theory of natu-ral selection; and Sigmund Freud promoted the unconscious world. These ideas, taken together, provided ammunition from the biological world, and seemed to be topped off by Einstein with his theory of rela-tivity and beliefs in a strictly deterministic world. As if that were not enough, along comes neuroscience with all sorts of findings that con-tinue to point us in that direction. The underlying contention is that free will is just happy talk. And just when you would think that the epicenter for such ideas is the physics department—after all, they got us into this mess—they are shaking their heads and have sneaked out the back door, along with many of the biologists, sociologists, and economists. The ones left sitting at the "hard" determinist table are the neuroscientists and Richard Dawkins, who said, "But doesn't a truly scientific, mechanistic view of the nervous system make nonsense of the very idea of responsibility?"[10] What happened? Why is the standard textbook understanding of determinism in trouble?

PHYSICS' DIRTY LITTLE SECRET

My son-in-law would say that the cause of the ball rolling across my floor is the floor isn't level. Then my three-year-old grandson would ask why it isn't level. Both Newton and my son-in-law would say I had made inaccurate measurements and then would point out that if my initial measurements had been more accurate, then my floor would be level. Defending myself by stretching a point, I could point out that because there is uncertainty in every measurement, the initial condi-tions could not be measured with complete accuracy. If the initial measurement is uncertain, then the results derived from that measure-ment are also uncertain. Maybe my floor would have been level, and maybe not. But Newton would have disagreed. Up until 1900, when a pesky Frenchman shook things up, physicists assumed that by making better and better initial measurements the uncertainties in the predictions would be less and that it was theoretically possible to obtain nearly perfect predictions for the behavior of any physical system. Well, of course, Newton would have been right about the physical universe as it pertains to my floor, but, as usual, things aren't so simple.

Chaos Theory

In 1900 Jules Henri Poincaré, a French mathematician and physicist, threw a fly in the ointment when he made a major contribution to what had become known as "the three-body problem" or "n-body problem" that had been bothering mathematicians since Newton's time. Newton's laws when applied to the motion of planets was completely deterministic, thus implying that if you knew the initial position and velocity of the planets, you could accurately determine their position and velocity in the future (or the past for that matter). The problem was that the initial measurement, no matter how carefully done, was not infinitely precise, but had a small degree of error. This didn't bother anyone very much because they thought the smaller the imprecision of the initial measurement, the smaller the imprecision of the predicted answer.

Poincaré found that while simple astronomical systems follow the rule that reducing the initial uncertainty always reduced the uncertainty of the final prediction, astronomical systems consisting of three or more orbiting bodies with interactions between all three did not. Au contraire! He found that even very tiny differences in initial measurements, *over time,* would grow at quite a clip, producing substantially different results, far out of proportion with what would be expected mathematically. He concluded that the only way to obtain accurate predictions for these complex systems of three or more astronomical bodies would be to have absolutely accurate measurements of the initial conditions, a theoretical impossibility. Otherwise, over time, any minuscule deviation from an absolutely precise measurement would result in a deterministic prediction with scarcely less uncertainty than if the prediction had been made randomly. In these types of systems, now known as chaotic systems, extreme sensitivity to initial conditions is called *dynamical instability* or *chaos,* and long-term mathematical predictions are no more accurate than random chance. So the problem with a chaotic system is that using the laws of physics to make precise long-term predictions is impossible, even in theory. Poincaré's work, however, simmered in the background for many decades until a weatherman got curious.

During the 1950s, mathematician-turned-meteorologist Edward Lorenz wasn't happy with the models that were being used for weather prediction (he had probably been blamed for too many ruined picnics). Weather depends on a number of factors such as temperature, humidity, airflow, and so on, and these are to a certain extent interdependent but nonlinear, that is, they are not directly proportional to one another. The models that were being used, however, were linear models. Over the course of the next few years he gathered data and began to put it together. He worked up a mathematical software program (which included twelve differential equations) to study a model of how an air current would rise and fall while being heated by the sun. One day, after obtaining some initial results from running his program, he decided that he would extend his calculations further. Because this was 1961, not only was his computer cumbersome, weighing in at 740 pounds, it was slow. He made a decision to restart the program in the middle of the calculation to save time, and this serendipitous lack of patience and his perceptive brain made him famous. After inputting the data the machine had calculated at that middle point in the previous run, he went out for coffee as the computer chugged along.

Lorentz expected he would get the same result as when he had last run the program—after all, computer code is deterministic. When he came back with his coffee, however, the results were completely different! No doubt exasperated, at first he thought it was a problem with the hardware, but eventually he traced it to the fact that instead of inputting the original number .506127, he had rounded off to the third decimal and only typed .506. Because Poincaré's chaotic systems had not seen the light of day for more than a half century, so small a difference was considered to be insignificant. For this system, however, a complex system with many variables, it wasn't! Lorenz had rediscovered chaos theory.

Weather is now understood to be a chaotic system. Long-term forecasts just are not feasible because there are too many variables that are impossible to measure with any degree of accuracy, and even if you could, the tiniest amount of imprecision in any one of the initial measurements would cause a tremendous variation in the end result. In 1972 Lorenz gave a talk about how even tiny uncertainties would eventually overwhelm any calculations and defeat the accuracy of a long-term forecast. This lecture, with the title, "Predictability: Does the Flap of a Butterfly's Wings in Brazil Set off a Tornado in Texas?" sired the term *butterfly effect*[11] and captured the imagination and fueled the fire of determinists. Chaos doesn't mean that the system is behaving randomly, it means that it is *unpredictable* because it has many variables, it is too complex to measure, and even if it could be measured, theoretically the measurement cannot be done accurately and the tiniest inaccuracy would change the end result an enormous amount. To determinists, it just means weather is a huge system with many variables but still follows deterministic behavior to such an extreme that something as minute as the flap of a butterfly's wing affects it.

Weather is an unstable system that exists far from thermodynamic equilibrium, as do most of nature's systems. These types of systems caught the eye of physical chemist Ilya Prigogine. As a child, Prigogine was drawn to archeology and music, and later as a university student he became interested in science. The comingling of these interests made Prigogine question Newtonian physics, which treated time as a reversible process. This didn't make sense to someone who had had these early interests in subjects where time proceeds in one direction. So while weather presented a problem for Newtonian physics because it is irreversible, it interested Prigogine. He called these types of systems "dissipative systems" and in 1977 won the Nobel Prize in Chemistry for the work that he pioneered on them. Dissipative systems do not exist in a vacuum but are thermodynamically open systems that exist in an environment where they are constantly sharing matter and energy with other systems. Hurricanes and cyclones are dissipative systems. They are characterized by the spontaneous appearance of symmetry breaking (emergence) and the formation of complex structures. Symmetry breaking is where small fluctuations acting on a system cross a critical point and determine which of several equally likely outcomes will occur. A well-known example is a ball sitting at the top of a symmetrical hill, where any disturbance will cause it to roll off in any direction, thus breaking its symmetry and causing a particular outcome. We will come back to this idea of emergence of complex systems in a bit.

So we now understand that weather forecasts can be accurate only in the short run. Even with the most super computer possible, long-term forecasts will always be no better than guesses. Well, hasn't it always been said that only a fool predicts the weather? And although weather is traditionally one of those safe topics to discuss, it may no longer be so at some dinner parties. If the presence of chaotic systems in nature, Poincaré's fly in the ointment, limits our ability to make accurate predictions with any degree of certainty using deterministic physical laws, it presents a quandary for physicists. It seems to imply that *either* randomness lurks at the core of any deterministic model of the universe *or* we will never be able to prove that deterministic laws apply in complex systems. Some physicists, because of this fact, are scratching their heads and thinking that it is meaningless to say that the universe is deterministic in its behavior. Now maybe at your house this is not a big deal, but imagine you are at a dinner party with, well, errr, how about Mr. Determinism himself, Baruch Spinoza, who said, "There is no mind absolute or free will, but the mind is determined for willing this or that by a cause which is determined in its turn by another cause, and this one again by another, and so on to infinity." Or maybe Albert Einstein, who said, "In human freedom in the philosophical sense I am definitely a disbeliever. Everybody acts not only under external compulsion but also in accordance with inner necessity." Hmmm, put a few other physicists in there and that would not be a good digestive environment. It turns out that Einstein was fighting his own determinism battles centered on quantum mechanics.

Quantum Mechanics Stirs Up a Hornet's Nest

During the five decades or so that chaos theory was simmering in the background, it was quantum mechanics that had grabbed the headlines, and most physicists were focused on the microscopic: atoms, molecules, and subatomic particles, not the balls in my living room or in Poincaré's sky. What they were finding out sent the world of physics into a tailspin. After three centuries, just when everyone was complacently assuming that Newton's laws were totally universal, they had found that atoms didn't obey the so-called universal laws of motion. How could Newton's laws be fundamental laws if the stuff of which objects are made, atoms, doesn't obey the same laws as the objects themselves? As Richard Feynman once pointed out, exceptions prove the rule . . . wrong.[12] Now what was going on? Atoms, molecules, and subatomic particles don't act like the balls in my living room. In fact, they are not balls at all, but waves! Waves of nothing! Particles are packets of energy with wavelike properties.

Crazy stuff happens in the quantum world. For instance, photons don't have mass, but angular momentum. Quantum theory was developed to explain why an electron stays in its orbit, which could neither be explained by Newton's laws nor Maxwell's laws of classical electromagnetism. It has successfully described particles and atoms in molecules, and its insights have led to transistors and lasers. But a philosophical problem lurks within quantum mechanics. Schrodinger's equation, which describes in a deterministic way how the wave function changes with time (and is reversible), cannot predict where the electron is in its orbit at any one state in time: that is a probability. If one actually measures the position, the act of measuring it distorts what the value would have been had it not been measured. This is because certain pairs of physical properties are related in such a manner that both cannot be known precisely at the same time: The more precisely one knows one property (by measuring it), the less precisely the other is known. In the case of the electron in orbit, the paired properties are position and momentum. If you measure the position, then it changes the momentum and vice versa. The theoretical physicist Werner Heisenberg presented this as the uncertainty principle. And uncertainty was not a happy thought for physicists and their determinist views but forced them to a different way of thinking. More than half a century ago, Niels Bohr, in his Gifford Lectures spanning 1948–1950, and even earlier in a 1937 article, was already pulling in the reins on determinism when he said, "The renunciation of the ideal of causality in atomic physics . . . has been forced upon us . . ."[13] and Heisenberg went even further when he said, "I believe that indeterminism, that is, is necessary, and not just consistently possible."[14]

Another lurking problem is the issue of time and causation. Time and semantics are two bugaboos that present themselves when you think about causation. When one recklessly and willy-nilly uses the word *causes* one can be thrown into an endless regression of questions and answers, as if being interviewed by a two-year-old who has just learned the word (with its attendant punctuation) *why*? Eventually in this regression of whys, as many determinists and reductions will point out, you will get down to atoms and subatomic particles. But this presents a fundamental problem, as systems theorist Howard Pattee, an emeritus professor at the State University of New York–Binghamton, points out:

[T]he microscopic equations of physics are time-symmetric and therefore conceptually reversible. Consequently the irreversible concept of causation is not formally supportable by microphysical laws, and if it is used at all it is a purely subjective linguistic interpretation of the laws. . . . Because of this time symmetry, systems described by such reversible dynamics cannot formally (syntactically) generate intrinsically irreversible properties such as measurement, records, memories, controls, or causes. . . . Consequently, no concept of causation, especially downward causation, can have much fundamental explanatory value at the level of microscopic physical laws.[15]

And for the semantic problem, Pattee adds, "[T]he concepts of causation have completely different meanings in statistical or deterministic models," and gives the following example: If you were to ask "What is the cause of temperature?" a determinist will assume that cause refers to a microscopic event and say it is caused by the molecules exchanging their kinetic energy by collisions. But the skeptical observer, scratching his head, will note that the measuring device averages this exchange, and does not measure the initial conditions of all the molecules and that averaging, my dear sir (or madam), is a statistical process. An average cannot be observable in a microscopic, determinist model. We have a case of apples and oranges. Pattee wags his finger at those who champion one model over the other and instead champions the idea that they are both needed and are complementary to each other. "I am using complementary here in Boltzmann's and Bohr's sense of logical irreducibility. That is, complementary models are formally incompatible but both necessary. One model cannot be derived from, or reduced to, the other. Chance cannot be derived from necessity, nor necessity from chance, but both concepts are necessary. . . . It is for this reason that our concept of a deterministic cause is different from our concept of a statistical cause. Determinism and chance arise from two formally complementary models of the world. We should also not waste time arguing whether the world itself is deterministic or stochastic since this is a metaphysical question that is not empirically decidable." I love that you get to tell everyone to hush when you are an emeritus professor.

Of course many determinists are anxious to point out that the chain of causes according to determinism is a chain of *events* not particles, so it never gets down to atoms or subatomic particles. Instead, it traces back to the big bang. In Aristotelian terms, the chain is a series of efficient causes rather than material causes.

Emergence

Smugly I point out to my son-in-law that the floor doesn't affect the atoms in the ball. Unfortunately, he is a voracious reader with an endlessly inquisitive mind. He points out that Newton's laws only seem to fail at the level of the atom, one of those things that the physicists with their super measuring devices stumbled upon. "We are not dealing with atoms, but balls. You are talking about another level of organization that doesn't apply here." The smart aleck brings up the topic of emergence. Emergence is when micro-level complex systems that are far from equilibrium (thus allowing for the amplification of random events) self-organize (creative, self-generated, adaptability-seeking behavior) into new structures, with new properties that previously did not exist, to form a new level of organization on the macro level.[16] There are two schools of thought on emergence. In *weak emergence,* the new properties arise as a result of the interactions at an elemental level and the emergent property is reducible to its individual components, that is, you can figure out the steps from one level to the next, which would be the deterministic view. Whereas, in *strong emergence,* the new property is irreducible, is more than the sum of its parts, and because of the amplification of random events, the laws cannot be predicted by an underlying fundamental theory or from an understanding of the laws of another level of organization. This is what the physicists stumbled upon, and they (and their left-brain interpreters) didn't like the inexplicable idea much, but many have come to accept that this is the way things are. Ilya Prigogine, however, was happy about one thing. He could identify the "arrow of time" as an emergent property that appears at a higher, macro, organizational level. Time does matter on the macro level as is obvious in biological systems. Emergence doesn't just apply to physics. It applies to all organized systems: cities emerge out of bricks, Beatlemania out of what? Calling a property emergent does not explain it or how it came to be, but rather puts it on the appropriate level to more adequately describe what is going on.

You may not know it, but authors do not have full jurisdiction over the titles of their books and the ultimate choice emerges (inexplicably?) from the publisher. I wanted to call my last book *Phase Shift*. A phase shift in matter, say from water to ice, is a change in the molecular organization resulting in different properties. I liked the analogy that the difference between the human brain and the brains of other animals is a change in the neuronal organization with resulting new properties. The publisher wasn't impressed. He called it *Human*. What has become obvious to most physicists (and apparently my son-in-law) is that at different levels of structure, there are different types of organization with completely different types of interactions governed by different laws, and one emerges from the other but does not emerge predictably. This is even true for something as basic as water turning to ice, as physicist Robert Laughlin has pointed out: Ice has so far been found to have eleven distinct crystalline phases, but none of them were predicted by first principles![17]

The balls in my living room are made up of atoms that behave as described by quantum mechanics, and when those microscopic atoms come together to form macroscopic balls, a new behavior emerges and that behavior is what Newton observed and described. It turns out that Newton's laws aren't fundamental, they are emergent; that is, they are what happens when quantum matter aggregates into macroscopic fluids and objects. It is a collective organizational phenomenon. The thing is, you can't predict Newton's laws from observing the behavior of atoms, nor the behavior of atoms from Newton's laws. New properties emerge that the precursors did not possess. This definitely throws a wrench into the reductionist's works and also throws a wrench into determinism. If you recall, the corollary to determinism was that every event, action, et cetera, are predetermined and can be predicted in advance (if all parameters are known). Even when the parameters of the atom are known, however, they cannot predict Newton's laws for objects. So far they can't predict which crystalline structure will occur when water freezes in different conditions.

So in some part because of chaos theory and perhaps more so because of quantum mechanics and emergence, physicists are sneaking out the determinism back door, with their tails between their legs. Richard Feynman, in his 1961 lectures to Caltech freshmen, famously declared: "Yes! Physics *has* given up. *We do not know how to predict what would happen in a given circumstance,* and we believe now that it is impossible—that the only thing that can be predicted is the probability of different events. It must be recognized that this is a retrenchment in our earlier ideal of understanding nature. It may be a backward step, but no one has seen a way to avoid it. . . . So at the present time we must limit ourselves to computing probabilities. We say 'at the present time,' but we suspect very strongly that it is something that will be with us forever—that it is impossible to beat that puzzle—that this is the way nature really is."[18]

The big question hovering over the head of the phenomenon of emergence is whether this unpredictability is a temporary state of affairs or not. Just because we don't know it yet, doesn't necessarily mean that it is unknowable, although it could be. Albert Einstein believed we considered things to be random merely out of ignorance of some basic property, whereas Niels Bohr believed that probability distributions were fundamental and irreducible. In some cases that have seemingly been explained, Adelphi University professor Jeffrey Goldstein, who studies complexity science, points out that it wasn't emergence that was the problem, but rather that the example used was not really an example of emergence, whereas in the case of a strange attractor,* "mathematical theorems support the inviolable unpredictability of this particular emergent. . . ." But as McGill University philosopher and physicist Mario Bunge points out, "Explained emergence is still emergence"[19] and even if one level can be ultimately derived from another "to dispense altogether with classical ideas seems sheer fantasy, because the classical properties, such as shape, viscosity, and temperature, are just as real as the quantum ones, such as spin and nonseparability. Shorter: the distinction between the quantum and classical levels is objective, not just a matter of levels of description and analysis."

Meanwhile back at the neuroscience department, however, hard determinism still reigns. Hard determinists have difficulty accepting that there is more than one level. They have a hard time accepting the possibility of the radical novelty that accompanies the emergence of a higher level. And why is this? It is because there is so much evidence that the brain functions automatically and that our conscious experience is an after-the-fact experience. At his point, let's once again remember what a brain is for. This is something that neuroscientists don't tend to think about much, but the brain is a decision-making device. It gathers information from all sorts of sources to make decisions from moment to moment. Information is gathered, computed, a decision is made, and *then* you get the sensation of conscious experience. Now you can actually do a little experiment for yourself that demonstrates that consciousness is a post hoc experience. Touch your nose with your finger and you will feel the sensation on your nose and your finger simultaneously. However, the neuron that carries the sensation from your nose to the processing area in the brain is only about three inches long, while the neuron from your hand is about three and a half feet long, and the nerve impulses travel at the same velocity.

---

* An attractor is a set (a collection of distinct objects) toward which a dynamical system evolves over time. A complicated set with a fractal structure is known as a strange attractor. (Wikipedia)

There is a difference of a few hundred (250–500) milliseconds in the amount of time that it takes for the two sensations to reach the brain, but you are not conscious of this time differential. The information is gathered from the sensory input and computed, a decision is made that both have been touched simultaneously even though the brain did not receive the impulses simultaneously, and only after that do you get the sensation of conscious experience. Consciousness takes time, but it arrives after the work is done!

CONSCIOUSNESS: A DAY LATE AND A DOLLAR SHORT

These time lapses have been documented repeatedly beginning more than twenty-five years ago. Benjamin Libet, a physiologist at the University of California–San Francisco, shook things up when he stimulated the brain of an awake patient during the course of a neurosurgical procedure and found that there was a time lapse between the stimulation of the cortical surface that represents the hand and when the patient was conscious of a sensation in the hand.[20] In later experiments, brain activity involved in the initiation of an action (pushing a button), occurred about five hundred milliseconds before the action, and that made sense. What was surprising was there was increasing brain activity related to the action, as many as three hundred milliseconds *before* the conscious intention to act, according to subject reports. The buildup of electrical charge within the brain that preceded what were considered conscious decisions was called *Bereitschaftspotential* or, more simply, readiness potential.[21]

Since the time of Libet's original experiments, as predicted by earlier psychologists, testing has become more sophisticated. Using fMRI, we now no longer think of the brain as a static system, but as a dynamic, ever-changing system that is constantly in action. Using these techniques, John-Dylan Haynes[22] and his colleagues expanded Libet's experiments in 2008 to show that the outcomes of an inclination can be encoded in brain activity up to ten seconds before it enters awareness! The brain has acted before its person is conscious of it. Not only that, from looking at the scan, they can make a prediction about what the person is going to do. The implications of this are rather staggering. If actions are initiated unconsciously, before we are aware of any desire to perform them, then the causal role of consciousness in volition is out of the loop: Conscious volition, the idea that you are willing an action to happen, is an illusion. But is this the right way to think about it? I am beginning to think not.

HARD DETERMINISTS: THE CAUSAL CLAIM CHAIN GANG

So the hard determinists in neuroscience make what I call the causal chain claim: (1) The brain enables the mind and the brain is a physical entity; (2) The physical world is determined, so our brains must also be determined; (3) If our brains are determined, and if the brain is the necessary and sufficient organ that enables the mind, then we are left with the belief that the thoughts that arise from our mind also are determined; (4) Thus, free will is an illusion, and we must revise our concepts of what it means to be personally responsible for our actions. Put differently, the concept of free will has no meaning. The concept of free will was an idea that arose before we knew all this stuff about how the brain works, and now we should get rid of it.

There is no disagreement among the neuroscientists about the first claim, that the brain enables the mind in some unknown way and the brain is a physical entity. Claim 2, however, has become a loose link and is under attack: Many physicists are no longer sure that the physical world is predictably determined because the nonlinear mathematics of complex systems does not allow exact predictions of future states. Now we have claim 3 (that our thoughts are determined) on shaky ground. Although some neuroscientists think we may prove that specific neuronal firing patterns will produce specific thoughts and that they are predetermined, none has a clue about what the deterministic rules would be for a nervous system in action. I think that we are facing the same conundrum that physicists dealt with when they assumed Newton's laws were universal. The laws are not universal to all levels of organization; it depends which level of organization you are describing, and new rules apply when higher levels emerge. Quantum mechanics are the rules for atoms, Newton's laws are the rules for objects, and one couldn't completely predict the other. So the question is whether we can take what we know from the micro level of neurophysiology about neurons and neurotransmitters and come up with a determinist model to predict conscious thoughts, the outcomes of brains, or psychology. Or even more problematic is the outcome with the encounter of three brains. Can we derive the macro story from the micro story? I do not think so.

I do not think that brain-state theorists, those neural reductionists who hold that every mental state is identical to some as-yet-undiscovered neural state, will ever be able to demonstrate it. I think conscious thought is an emergent property. That doesn't explain it; it simply recognizes its reality or level of abstraction, like what happens when software and hardware interact, that mind is a somewhat independent property of brain while simultaneously being wholly dependent upon it. I do not think it possible to build a complete model of mental function from the bottom up. If you do think this is possible, oddly enough, a spiny crustacean and a biologist have given us all pause on how it might all work.

THE SPINY LOBSTER PROBLEM

Eve Marder has been studying the simple nervous system and the resulting motility patterns of spiny lobster guts. She has isolated the entire pattern of the network with every single neuron and synapse worked out, and she models the synapse dynamics to the level of neurotransmitter effects. Deterministically speaking, from knowing and mapping all this information, she should be able to piece it together and describe the resulting function of the lobster gut. Her laboratory simulated more than 20 million possible network combinations of synapse strengths and neuron properties for this simple little nervous system.[23] By modeling all those combinations, it turned out that about 1–2 percent could lead to appropriate dynamics that would create the motility pattern observed in nature. Even though it is a small percent, it still turns out to be 100,000 to 200,000 different tunings that will result in the exact same behavior at any given moment (and this is a very simple system with few parts)! The philosophical concept of multiple realizability—the idea that there are many ways to implement a system to produce one behavior—is alive and well in the nervous system.

The enormous diversity of network configurations that could lead to an identical behavior leads one to wonder if it is possible to figure out, with single-unit analysis and very molecular approaches, what is going on to produce a behavior. This is a profound problem for the neuroscientist reductionist, because it shows that analyzing nerve circuits may be able to inform how the thing could work but not how it actually does work. On the surface, it seems to reveal how hard it is going to be to get a specific neuroscientific account of a specific behavior. Her work almost comes off as supporting the idea of emergence—that studying neurons won't get us to the right level of explanation. There are too many different states that can lead to one outcome. Should neuroscientists despair?

John Doyle doesn't think so and sees no need for that kind of talk at all. He points out that when considering multiple components of anything, it simply follows that as the number of circuit components and parameters grows there is a more than exponentially hugely growing set of possible circuits. Further, there is a smaller but still exponentially huge growing set of functional circuits. Importantly, the functional set is an exponentially vanishing fraction of the whole set. So even though the possible combinations are huge, the actual number of functional combinations is only a small percentage of that huge number.

Well, that is what Eve Marder and her colleagues discovered, and those relationships hold over many kinds of things not just lobsters. For example, as Doyle says, "there are a huge number of English words, something like more than 10^5 words. But take the word *organized*. . . . It has 9 different letters, so there are 362,880 sequences with just those nine letters, but only one of them is a functional English word. So any long random string of letters is vanishingly unlikely to be a real word (e.g., *roaginezd*), yet there are still a huge number of words." As Doyle points out, this is a good thing, because it is consistent with the idea that the brain is a layered system. Being layered buys a lot. It gets down to the idea of robustness. The layer below creates a very robust yet flexible platform for the emergent layer above.

Marder's work has revealed the problem for neuroscientists. The task is to further understand how the various layers of the brain interact, indeed how to even think about it and develop concepts and a vocabulary for those interdependent interactions. Working from this perspective has the possibility of not only demystifying what is truly meant by concepts such as emergence, but also allows for insights on how layers actually communicate with one another.

Even if we assume that claim 3—thoughts arising from our minds are determined—is true, then we are led to claim 4, that free will is an illusion. Putting aside the long history of compatibilism—the idea that people are free to choose an idea, more or less by assertion, in a determined universe—what does it really mean to talk about free will? "Ah, well, we want to be free to make our own decisions." Yes, but what do we want to be free from? We don't want to be free from our experience of life, we need that for our decisions. We don't want to be free from our temperament because that also guides our decisions. We actually don't want to be free from causation, we use that for predictions. A receiver trying to catch a football does not want to be free from all the automatic adjustments that his body is making to maintain his speed and trajectory as he dodges tackles. We don't want to be free from our successfully evolved decision-making device. What do we want to be free from? This topic draws quite a bit of attention, as you can well imagine. I would like, however, to talk about the system in a different light.

YOU'D NEVER PREDICT THE TANGO IF YOU ONLY STUDIED NEURONS

For literally thousands of years, philosophers and nearly everyone else have argued about whether the mind and body are one entity or two. The belief that people are more than just a body, that there is an essence, a spirit or mind, whatever it is that makes you "you" or me "me," is called *dualism*. Descartes is perhaps most famous for his dualistic stance. The idea that we have an essence beyond our physical selves comes so easily to us that we would think it odd if you were to resort to a mere physical description to describe someone. A friend of mine who recently met the retired Supreme Court justice Sandra Day O'Connor did not describe her height, hair color, or age, but said, "She is spunky and sharp as a tack." She described her mental essence. While determinism has supplanted dualism in the brain sciences, it falls short of explaining behavior and our sense of personal responsibility and freedom.

I think that we neuroscientists are looking at these capacities from the wrong organizational level. We are looking at them from the individual brain level, but they are emergent properties found in the group interactions of many brains. Mario Bunge makes a point that we neuroscientists should heed: "[W]e must place the thing of interest in its context instead of treating it as a solitary individual." The idea, which was difficult for physicists to swallow, but swallow most of them have, is that something happens that can't be captured from a bottom-up approach. Reductionism in the physical sciences has been challenged by the principle of emergence. The whole system acquires qualitatively new properties that cannot be predicted from the simple addition of those of its individual components. One might apply the aphorism that the new system is greater than the sum of its parts. There is a phase shift, a change in the organizational structure, going from one scale to the next. Why do we believe in this sense of freedom and personal responsibility? "The reason we believe them, as with most emergent things, is because we observe them." Although the physicist Robert Laughlin was commenting about phase transitions such as changing from water to ice, he may as well have been talking about our feelings of responsibility and freedom.

In speaking about the phenomenon of emergence in 1972, Nobel Prize–winning physicist Philip W. Anderson in his seminal paper, "More Is Different," reiterated the idea that we can't get the macro story from the micro story: "The main fallacy in this kind of thinking is that the reductionist hypothesis does not by any means imply a 'constructionist' one: The ability to reduce everything to simple fundamental laws does not imply the ability to start from those laws and reconstruct the universe. In fact, the more the elementary particle physicists tell us about the nature of the fundamental laws, the less relevance they seem to have to the very real problems of the rest of science, much less to those of society."[24] He later waggles his finger at biologists, and no doubt at us neuroscientists, too, "The arrogance of the particle physicist and his intensive research may be behind us (the discoverer of the positron said 'the rest is chemistry'), but we have yet to recover from that of some molecular biologists, who seem determined to try to reduce everything about the human organism to 'only' chemistry, from the common cold and all mental disease to the religious instinct. Surely there are more levels of organization between human ethology and DNA than there are between DNA and quantum electrodynamics, and each level can require a whole new conceptual structure."

In his wonderful book *A Different Universe,* Robert Laughlin, who won the Nobel Prize in Physics in 1998, said about the dawning of the understanding of emergence, "What we are seeing is a transformation of worldview in which the objective of understanding nature by breaking it down into ever smaller parts is supplanted by the objective of understanding how nature organizes itself."

Physicists have realized that a complete theoretical understanding of the microscopic constituents does not suggest a new set of general theories for how they are put together into interesting macromolecular structures and how the processes work that make it what it is. That nature does it is in no way in question, but whether we can theorize, predict, or understand this process is to Richard Feynman highly improbable, and Philip Anderson and Robert Laughlin believe it is impossible. The upwardly causal constructionist view that understanding the nervous system will allow us to understand all the rest of it is not the way to think about the problem.

Emergence is a common phenomenon that is accepted in physics, biology, chemistry, sociology, and even art. When a physical system does not demonstrate all the symmetries of the laws by which it is governed, we say that these symmetries are spontaneously broken. Emergence, this idea of symmetry breaking, is simple: Matter collectively and spontaneously acquires a property or preference not present in the underlying rules themselves. The classic example from biology is the huge, towerlike structure that is built by some ant and termite species. These structures only emerge when the ant colony reaches a certain size (more is different) and could never be predicted by studying the behavior of single insects in small colonies.

Yet, emergence is mightily resisted by many neuroscientists, who sit grimly in the corner and continue to shake their heads. They have been celebrating that they have finally dislodged the homunculus out of the brain. They have defeated dualism. All the ghosts in the machine have been banished and they, as sure as shootin', are not letting any back in. They are afraid that to put emergence in the equation may imply that something other than the brain is doing the work and that would let the ghost back into the deterministic machine that the brain is. No emergence for them, thank you! I think this is the wrong way for neuroscientists to look at the problem. Emergence is not a mystical ghost but the going from one level of organization to another. You, alone on the proverbial desert island, or for that matter, alone in your house on a rainy Sunday afternoon, follow a different set of rules than you do at a cocktail party at your boss's house.

The key to understanding emergence is to understand that there are different levels of organization. My favorite analogy is that of the car, which I have mentioned before. If you look at an isolated car part, such as a cam shaft, you cannot predict that the freeway will be full of traffic at 5:15 P.M. Monday through Friday. In fact, you could not even predict the phenomenon of traffic would ever occur if you just looked at a brake pad. You cannot analyze traffic at the level of car parts. Did the guy who invented the wheel ever visualize the 405 in Los Angeles on Friday evening? You cannot even analyze traffic at the level of the individual car. When you get a bunch of cars and drivers together, with the variables of location, time, weather, and society, all in the mix, then at *that* level you can predict traffic. A new set of laws emerge that aren't predicted from the parts alone.

The same holds true for brains. Brains are automatic machines following decision pathways, but analyzing single brains in isolation cannot illuminate the capacity of responsibility. Responsibility is a dimension of life that comes from social exchange, and social exchange requires more than one brain. When more than one brain interacts, new and unpredictable things begin to emerge, establishing a new set of rules. Two of the properties that are acquired in this new set of rules that weren't previously present are responsibility and freedom. They are not found in the brain, just as John Locke declared when he said, "the will in truth, signifies nothing but a power, or ability, to prefer or choose. And when the will, under the name of a faculty, is considered, as it is, barely as an ability to do something, the absurdity in saying it is free, or not free, will easily discover itself."[25] Responsibility and freedom are found, however, in the space between brains, in the interactions between people.

HOW TO RILE A NEUROSCIENTIST

Modern neuroscience is happy to accept that human behavior is the product of a probabilistically determined system, which is guided by experience. But how is that experience doing the guiding? If the brain is a decision-making device and is gathering information to inform those decisions, then can a mental state that is the result of some experience or the result of some social interaction affect or constrain future mental states? If we all were French we would, in exasperation, jut out our upper lip and let out an expiration, shrug, and say, "but of course," unless you were a neuroscientist or perhaps a philosopher. This means top-down causation. Suggesting top-down causation to a group of neuroscientists are fightin' words. It is to your peril to invite a group of them to your house and bring it up at dinner. Better we invite the physicist Mario Bunge, who will tell us that we "should supplement every bottom-up analysis with a top-down analysis, because the whole constrains the parts: just think of the strains in a component of a metallic structure, or the stress in a member of a social system, by virtue of their interactions with other constituents of the same system."

If we invite our systems control expert, Howard Pattee, he will be happy to tell us that while causation has no explanatory value at the level of physical laws, it certainly does at higher levels of organization. For instance, it is helpful to know that iron deficiency causes anemia. Pattee suggests that the everyday meaning of causation is pragmatic and is used for events that are controllable. Controlling the iron level will fix the anemia. We cannot change the laws of physics, but we can change the iron level. When a car rear-ends another car at the bottom of a hill, we say that the accident was caused by the worn-out brakes, something we can point our finger at and control. We do not, however, blame the laws of physics or all the chance circumstances that are not under our control (the fact that there was another car stopped at the light at the bottom of the hill, all the reasons that led to that driver being there, the timing of the traffic lights, and so on). Pattee sees this tendency to identify a single controllable cause "that, by itself, might have prevented the accident but maintained all other expected outcomes," rather than as the result of a complex system as "one reason that downward causation is problematic. In other words, we think of causes in terms of the simplest proximal control structures in what would otherwise turn into an endless chain or network of concurrent, distributed causes." That is to say, downward causation is chaotic and unpredictable.

And where does control enter into the picture, Pattee asks? Not at the micro level, because by definition physical laws describe only those relations between events which don't vary from one observer to the next. When a parent sternly asks, "Why did you cheat on your test?" and they receive the answer, "It was just atoms following the laws of physics," which is the universal cause for all events, the child will be labeled a smart aleck and be duly punished, probably even by the most reductionist of parents. The kid's explanation needs to come up a few levels of behavior to where control can be exerted. Control implies some form of constraint. Control is not eating the jelly donut because you know it is not healthy, and not cheating on the test because, well, if you get caught you get in some kind of trouble. Control is an emergent property.

In neuroscience when you talk about downward causation you are suggesting that a mental state affects a physical state. You are suggesting that a thought at the Macro A level can affect the neurons on the Micro B physical level. The first question is, how do we get from the level of neurons (Micro B) to the emergent thought (Macro A)? David Krakauer, a theoretical biologist at the Santa Fe Institute, emphasizes that "the trick, for any level of analysis, is to find the effective variables containing all the information from below required to generate all the behavior of interest above. This is as much an art as a science. Now, 'bottom-up causality' (going from a B Micro level, a neuron, to an A macro level, a thought) can be both intractable and incomprehensible. 'Top-down causality' refers to the description of Macro A causing Micro B when A is expressed in higher-level effective variables and dynamics, and B in terms of the microscopic dynamics. Physically, all the interactions are microscopic (B–B) but not all the microscopic degrees of freedom matter."[26] That is, B can generate A, but A is still made up of B.

For example, Krakauer points out that when we program a computer, or control the computer in Pattee's world, "we interface with a complex physical system that performs computational work. We do not program at the level of electrons, Micro B, but at a level of a higher effective theory, Macro A (for example, Lisp programming) that is then compiled down, without loss of information, into the microscopic physics. Thus, A causes B. Of course, A is physically made from B, and all the steps of the compilation are just B with B physics. But from our perspective, we can view some collective B behavior in terms of A processes."

If we go back to my living room, the atoms come together and can generate the ball rolling across the floor, but the ball is still made up of atoms. We view the collective behavior of the atoms, Micro B, at the higher organizational level of the ball, Macro A, and we see it doing ball behavior following Newton's laws, but the atoms are there at the core doing their own thing and following a different set of laws. In brain science we use concepts like anger, tone, and perspective for our Macro A states. These are the A coarse-grained variable states that we view standing in for the B micro states. Krakauer continues: "We work well with the A level, due to limitation of our own introspective awareness. Internally, something does the compiling before it reaches consciousness. So maybe either A or the compiler can be thought of as a language of thought. We are not separate from the machine, that Micro B level, but we understand ourselves at suitable A levels.

"The deeper point is that without these higher levels, there would be no possibility of communication, as we would have to specify every particle we wish to move in the utterance, rather than have the mind-compiler do the work." There is an absolute necessity for emergence to occur to control this teeming, seething system that is going on at another level. The overall idea is that we have a variety of hierarchical emerging systems erupting from the level of particle physics to atomic physics to chemistry to biochemistry, to cell biology to physiology emerging into mental processes.

COMPLEMENTARITY *SI*, DOWNWARD CAUSATION *NO*

Once a mental state exists, is there downward causation? Can a thought constrain the very brain that produced it? Does the whole constrain its parts? This is the million-dollar question in this business. The classic puzzle is usually put this way: There is a physical state, P1, at time 1, which produces a mental state, M1. Then after a bit of time, now time 2, there is another physical state, P2, which produces another mental state, M2. How do we get from M1 to M2? This is the conundrum. We know that mental states are produced from processes in the brain so that M1 does not directly generate M2 without involving the brain. If we just go from P1 to P2 then to M2, then our mental life is doing no work and we are truly just along for the ride. No one really likes that notion. The tough question is, does M1, in some downward-constraining process, guide P2, thus affecting M2?

We may get a little help with this question from geneticists. They used to think gene replication was a simple, upwardly causal system: Genes were like beads on a string that make up a chromosome that replicates and produces identical copies of itself. Now they know that genes are not that simple, that there is multiplicity of events going on. Our systems control guy, Howard Pattee, finds that a good example of upward and downward causation is the genotype-phenotype mapping of description to construction. It "requires the gene to describe the sequence of parts forming enzymes, and that description, in turn, requires the enzymes to read the description. . . . In its simplest logical form, the parts represented by symbols (codons) are, in part, controlling the construction of the whole (enzymes), but the whole is, in part, controlling the identification of the parts (translation) and the construction itself (protein synthesis)." And once again Pattee wags his finger at extreme positions that champion which is more important, upward or downward. They are complementary.

It is this sort of analysis that finds me realizing the reasoning trap we can all too easily fall into when we look to Benjamin Libet's kind of fact, that the brain does something before we are consciously aware of it. With the arrow of time all moving in one direction, with the notion that everything is caused by something before it, we lose a grip on the concept of complementarity. What difference does it make if brain activity goes on before we are consciously aware of something? Consciousness is its own abstraction on its own time scale and that time scale is current with respect to it. Thus, Libet's thinking is not correct. That is not where the action is, any more than a transistor is where the software action is.

Setting a course of action is automatic, deterministic, modularized, and driven not by one physical system at any one time but by hundreds, thousands, and perhaps millions. The course of action taken appears to us as a matter of choice, but the fact is, it is the result of a particular emergent mental state being selected by the complex interacting surrounding milieu.[27] Action is made up of complementary components arising from within and without. That is how the machine (brain) works. Thus, the idea of downward causation may be confusing our understanding. As John Doyle says, "Where is the cause?" What is going on is the match between ever-present multiple mental states and the impinging contextual forces within which it functions. Our interpreter then claims we freely made a choice.

It gets more complicated. We are now going to have to consider the social context and the social constraints on individual actions. There is something going on at the group level.

Chapter Five

THE SOCIAL MIND

IF YOU PICK UP A YOUNG BABY AND STICK YOUR TONGUE out at her, at some point she will stick her tongue out at you. It is as if you two are having a nice little social interaction. Her behavior is not learned. She appears to be automatically imitating your action and as a result appears to be engaging you socially. You may not think this is high-level communication, but maybe it is. When this clicks in, a baby has looked at you, recognized you as imitable (that is, as an animate object rather than a lamp), saw your tongue, recognized that she has a tongue, figured out, from of all her muscles that she has control of, which one is the tongue, and out it goes! She is a baby! How does she know a tongue is a tongue—or does she? How does she know how to use the neural system that is in charge of the tongue and move it? Why does she even bother doing it?

Babies first enter the social world through imitation. They understand they are like other people and imitate human actions, but not those of objects.[1] This is because the human brain has specific neural circuits for identifying biological motion and inanimate object motion, along with specific circuits to identify faces and facial movement.[2] A baby cannot do much to enter the social world and form a link with another person before she can sit up, control her head, or talk. But she can imitate. When you hold a baby, what links the two of you together in the social world are her imitative actions. She doesn't just lie there like a lump of lead but responds in a way that you can relate to.

In the last chapter, I left off suggesting that responsibility arises out of social interaction and that the mind constrains the brain. We are now going to see how we incorporate social dynamics into personal choice, how we figure out the intentions, emotions, and goals of others in order to survive, and understand how social process constrains individual minds. Thinking that individuals are constrained by social process is a rather irksome topic to Americans. After all, we are a country that has favored rugged individuality, told a whole generation to strike out on their own with the headlines, "GO WEST, YOUNG MAN, GO WEST!" and made an icon of the lone cowboy. It was reported that when Henry Ford was told "Mr. Ford, a man, Charles Lindbergh just flew over the Atlantic Ocean by himself" he replied, "That's nothing. Tell me when a committee flies over." Our individualist thinking has actually influenced how we have approached and what we have focused on while studying humans and brain function. Thus, we do know a lot about the psychology of the individual, but we are just now understanding the neuroscience of the influences of social interactions.

STANDARD EQUIPMENT: BORN TO BE SOCIAL

It turns out that we are wired from birth for social interactions. A great many of our social abilities come hardwired from the baby factory. The advantage of hardwired abilities, of course, is they work immediately and don't have to be learned, as opposed to all of the survival skills that do. David and Ann Premack got the ball (or I should say triangle) rolling on the studies of intuitive social skills by looking for what, if any, social concepts toddlers understood. It had been shown in the early 1940s that, when presented with films of geometric shapes moving in ways that suggest intention or goal-directed behavior (moving in ways that an animal would move), people will even attribute desires and intentions to geometric figures.[3] The Premacks demonstrated that even ten- to fourteen-month-old infants, watching objects that appeared to be self-propelled and goal-oriented, automatically interpreted the objects as intentional, and, more important, they assigned a positive or negative value to the interaction between intentional objects.[4] This work was extended by J. Kiley Hamlin, Karen Wynn, and Paul Bloom, who showed that even six- to ten-month-old infants evaluated others based on their social behavior. These infants watched a video in which an animated triangle with eyes tries to make it up a hill, and it is either helped by a push from a circle or hindered by a push from a square. After the video, the babies were given a choice of the circle or the square on a tray, and they grabbed the "helper" circle.[5] The ability to evaluate other people is essential for navigating the social world. It appears that even preverbal infants can figure out who is helpful and who is not, an obvious advantage to a child who needs many years of help to survive.

Looking for early signs of helping behavior in the children themselves, Felix Warneken and Michael Tomasello believe that even children as young as fourteen months old will altruistically help another. Without encouragement or praise, they will pick up an object that someone has accidentally dropped and hand it back to them,[6] sometimes, even if they must stop doing an activity that they are enjoying.[7] This, of course, involves not only understanding that others have goals and what they are, but also altruistic behavior to non-kin, an evolutionarily rare behavior that could have foundations in our chimp relatives and already manifests itself in fourteen-month-old children.[8] Helping appears to be something that comes naturally and is not something that is exclusively learned. Unlike chimpanzees, however, other research from Tomasello's lab found that twelve-month-old children will also freely give information. If they know where an object is that someone is looking for, they will point to it.[9]

Interestingly, altruistic behavior, which is appearing to be innate in humans, is influenced by social experience and cultural transmission.[10] Young children by age three begin to *inhibit* some of their naturally altruistic behavior. They become more discriminating about whom they help. They share more often with others who have shared with them in the past.[11] Chimps do the same thing,[12] exhibiting at least some of the characteristics of reciprocal altruism. Social norms and rules also begin to influence altruistic behavior in preschool children.[13]

ORIGINS OF SOCIAL BEHAVIOR: SAFETY IN NUMBERS

How did such social behavior evolve? When I think about the evolution of human social processes, I divide it into two stages. Evolutionary psychologists continually remind us to remember the environment that our ancestors were living in, which was very sparsely populated. Even as recently as 10,000 B.C., when the glacial ice of the last ice age was retreating in North America, people were few and far between. As the early hominids banded together in small groups for protection from predators and help in hunting, social adaptations were being evolved. For most of human history, food sources were widely scattered, and these small groups were nomadic. It has not been until very recently that the population has become dense, which all started with the development of agriculture and the change to a sedentary lifestyle. In fact, the number of people who were alive in 1950 was about equivalent to the number who had been alive for the entire previous history of the world.

As population density increased, the second stage kicked in: adaptations for navigating and managing the increasingly populated social world. There are now 6.7 billion of us, more than twice the population of 1950. The amazing thing is that we as a species are becoming less violent and get along rather well, contrary to what you may hear on the evening news. The troublemakers, although still very much of a problem, are actually few and far between, perhaps five percent of the population. As a species, we don't like to kill, cheat, steal, and be abusive. This fact brings us to think about our social interactions and how our mental life is codependent upon others. How do we recognize the emotional states of others in order to understand them, and how do we come by the moral and social rules that we live by? Are the rules learned, innate, or both? What abilities do we have to navigate all the social interactions that we daily face and how did they come about? Are we rational beings living by a set of personal rules, or are there group dynamics that can hijack us? Alone in a situation, does someone act the same as if they were in a group?

IT TAKES TWO TO TANGO

The realization has come slowly to neuroscience and psychology that we can't just look at the behavior of one brain. Asif Ghazanfar, who studies vocalization in both macaques and humans at Princeton University, makes the point that not only is there a dynamic relationship going on that involves different parts of the brain, but there is also a dynamic relationship with the other animal that is being listened to. The vocalizations of one monkey modulate the brain processes going on in the other monkey. This is true for humans also. Uri Hasson at Princeton measured the brain activity of a pair of conversing subjects with fMRI. He found that the listener's brain activity mirrored the speaker's; some areas of the brain even showed predictive anticipatory responses. When there were such anticipatory responses, there resulted greater understanding.[14] The behavior of one person can affect another person's behavior. The point is that we now understand that we have to look at the whole picture, not just one brain in isolation if we are to understand a more full set of forces in play.

This was a concept that dawned on primatologists many years ago. In 1966, Alison Jolly concluded a paper about Lemur social behavior with, "It thus seems likely that the rudiments of primate society preceded the growth of primate intelligence, made it possible, and determined its nature."[15] The overall line of reasoning goes something like the following, which I have reviewed in my book *Human*.

BIG BRAINS AND COMPETITION, OR THE ORIGIN OF PARTY SCHOOLS

There have been many theories about what forces were relentlessly driving the enlarging of the human brain. Through the processes of natural and sexual selection, it is coming to be accepted that two main factors were at work: A diet with sufficient calories to feed the increasingly metabolically expensive bigger brain, and the challenge of living in a large group (that "social world," which was necessary to guard against predators and to hunt and gather food). Banding together in social groups resulted in its own set of problems, including competition with others for limited resources, both food and prospective mates. Alison Jolly's observation, followed by those of others, led Richard Byrne and Andrew Whiten at the University of St. Andrews, Scotland, to propose what has come to be known as the social brain hypothesis. They proposed that primates have more complex social skills than nonprimates and that living in complexly bonded social groups is more challenging than dealing with the physical world. (Everyone knows it is easier to fix the toaster at the back of the shop than doing customer service up front.) The cognitive challenge of surviving in increasingly larger social groups selected for increases in brain size and function.[16]

Most monkeys and apes live in long-lasting groups, so that familiar conspecifics are major competitors for access to resources. This situation favors individuals who can offset the costs of competition by using manipulative tactics, and skillful manipulation depends on extensive social knowledge. Because competitive advantage operates relative to the ability of others in the population, an "arms race" of increasing so-cial skills results, which is eventually brought into equilibrium by the high metabolic cost of brain tissue.[17]

To be successful in a social group involves more than competition. One must also cooperate; otherwise such activities as joint hunting wouldn't work. To address this issue, developmental and comparative psychologists Henrike Moll and Michael Tomasello have suggested the Vygotskian intelligence hypothesis, named after Lev Vygotsky, an early twentieth-century Russian psychologist.* They propose that while cog-nition in general was driven mainly by social competition, other as-pects of cognition that they consider to be unique to humans (the cognitive skills of shared goals, joint attention, joint intentions, and cooperative communication), were driven by or were constituted of social cooperation, which is needed to create such things as complex technologies, cultural institutions, and systems of symbols, and not by social competition.[18]

THE BIGGER THE PARTY, THE BIGGER THE BRAIN

Oxford University anthropologist Robin Dunbar has provided support for some type of social component driving the evolutionary expansion of the brain. He has found that each primate species tends to have a typi-cal social group size; that brain size correlates with social group size in primates and apes; that the bigger the neocortex, the larger the social group; and that the great apes require a bigger neocortex per given group size than do the other primates.[19] While a typical social group size for a chimpanzee is about 55 individuals, Dunbar predicted from

* Vygotsky investigated how child development and learning was guided by so-cial interactions with parents and others, through which the child learns the cultural habits of mind, speech patterns, written language, and symbols.

the size of human brains that the typical social group size for humans is about 150 individuals.[20] Then he studied actual human social groups, and it turns out that this social group size has been constant for humans from prehistoric times through today. Not only was this the size of related groups that gathered together once a year for traditional ceremonies in ancestral hunter-gatherers, it is also the size of modern-day hunter-gatherer societies and modern-day Christmas card lists in personal address books.[21] Today's social networking sites appear to be no different. In an ongoing study, Dunbar has so far found that even people with hundreds of "friends" interact with a limited number of them. "The interesting thing is that you can have 1,500 friends but when you actually look at traffic on sites, you see people maintain the same inner circle of around 150 people that we observe in the real world."[22]

Research has shown that 150–200 people are the number of people that can be controlled without an organizational hierarchy.[23] It is the number of people one can keep track of, maintain a stable social relationship with, and would be willing to help with a favor. Yet, why is our social group size limited? To have social relationships, you call on five cognitive abilities: (1) you must interpret visual information to recognize others, then (2) be able to remember both faces and (3) who has a relationship with whom; (4) you must process emotional information, and then (5) manipulate information about a set of relationships. Dunbar has found that it is the ability to manipulate information about a set of relationships that is the limiting factor. The other processes are not running at capacity. Information about social relationships requires additional processing capacity, as well as specific specializations, while the others do not.

WANDERLUST LOST

Since a myriad of forces drives evolution, one has to be careful not to focus too much on just one aspect. Years ago I was privileged to be part of a small study group that Leon Festinger had put together that also included David Premack and social psychologist Stanley Schachter. Leon was interested in what could account for the vast differences between our species and other animals. He pointed out that one of the possible consequences of social behavior, which triggered so many changes, was becoming sedentary and abandoning the nomadic lifestyle. Between 10,500 and 8,500 B.C., many things that had been accumulating over the past thousands of years came together and made a major change in lifestyle possible. There was the end of the last glacial period; there was control of fire and more effective hunting; the dog had been domesticated (the social world really took off, now that man had a best friend!); there was an increased consumption of fish and a greater reliance on storable cereal grains. Festinger concluded that sedentary existence was the fundamental change that irreversibly altered the course of human evolution. A sedentary lifestyle allowed humans to reproduce more successfully (owing to a reduction of miscarriages and a reduction of spacing between children), and group size quickly increased to around 150. Although the environment and natural resources normally temper the population increases caused by the endogenous drive to reproduce, this was not so for humans. They were able, sooner or later, to find or invent solutions to problems and markedly change their environment while they were evolving. So as sedentary groups formed, their populations increased; around 7,000 B.C. someone had a big idea, and agriculture came on the scene. This was followed by increasing specialization from 6,000 to 4,500 B.C., which required more interdependence in communities, which in turn created a greater potential for status and power differences. Meanwhile, there was the development of natural and religious technologies, social rules, gossip, and moral stance to control and organize these communities of people.

YOU CAN'T KEEP 'EM
DOWN ON THE FARM . . .

The point is that along with all our automatic processes, there is this whole living environment changing and impacting our behavior, thinking, and perhaps even our genome. Primitive social behavior was largely intact before the sedentary life style developed. It was sedentary life and the subsequent civilizations, however, which provided the milieu in which complex social behaviors arose and the social brain flourished. We entered into what I call stage 2, a coevolution with emerging civilization that continues to shape social components of the human brain, even today.

Coevolution?

How could such coevolution come about? In essence, natural selection is a case of downward causation with a sort of feedback mechanism to a passive selector. The environment is downwardly causal in that whatever survives, survives the effects of the environment for whatever reason. The survivor is the feedback, in that he reproduces and allows the next generation in its turn to be acted upon by the environment. Now if the survivor slightly changes the environment, then what the slightly changed environment selects may be slightly changed. It may be no different with social processes: The social environment is just another factor contributing to the overall environment that is selecting in a downwardly causal way, with a feedback mechanism at work.

As I mentioned before, a genetically fixed trait is always superior to one that must be learned because learning may or may not happen. Time, energy, and opportunity to learn are required and may not be available. For both an infant and an adult, hard wired automatic responses offer a survival advantage, but as one progresses through life, flexibility in the face of change is also advantageous. The physical environment is not stable. Earthquakes, volcanic eruptions, ice ages, droughts, famines, and so on, do happen. Change and the unexpected do arise. As philosopher David Papineau points out, "As a general rule, then, we can expect that genetic fixity will be favoured when there is long-term environmental stability, and that learning will be selected for when there are variable environments. Given environmental stability, genetic fixity will have the . . . advantages of reliable and cheap acquisition. But these advantages can easily be outweighed by loss of flexibility when there is significant environmental instability."[24] The social environment can also be unstable, as evidenced by marked changes in population and its geographical distribution.

In 1896, the American psychologist James Mark Baldwin, working within the framework of Darwinian selection, sought a way to explain the evolution of traits that were not fixed but learned during an organism's life. At first glance this sounds like Lamarckian genetics, the inheritance of acquired characteristics, but it is not. He came up with the idea that while acquired traits cannot be inherited, the *tendency* to acquire certain traits can be.[25] (Using my old example, one has the tendency to acquire the fear of snakes but does not have the tendency to acquire the fear of flowers.) The first mention of the Baldwin effect at the Gifford lectures was by Conrad Waddington in 1971. In essence, the Baldwin effect is a mechanism that explains the evolution of phenotypic (observable trait) plasticity, the ability which allows an organism to be flexible in adapting its behavior to changing environments. As evolutionary neurobiologists Leah Krubitzer and Jon Kaas put it,

> Although the phenotype generated is context-dependent, the ability to respond to the context has a genetic basis. . . . In essence, the Baldwin effect is the evolution of the ability to respond optimally to a particular environment. Thus, genes for plasticity evolve, rather than genes for a particular phenotypic characteristic, although selection acts upon the phenotype.[26]

Becoming Flexible Isn't from Doing Yoga

There are two types of biological mechanisms that can result in the Baldwin effect: genetic assimilation and niche construction. Krubitzer and Kaas explain genetic assimilation:

> A particular phenotypic characteristic that is optimal for a given environment can become incorporated into the genome over successive generations by endowing a selective advantage to those individuals who display these optimal characteristics, and who have a strong correlation between genotypic and phenotypic space. This characteristic is then displayed even in the absence of the environmental condition that first produced it. This process, known as *genetic assimilation* [italics added], accounts for how activity-dependent modifications to the phenotype come under genetic control and become part of the evolutionary process.

The other biological mechanism is niche construction. Hidden in plain view, niche construction[27] has been a neglected topic in evolutionary theory until recently. F. John Odling-Smee, Kevin N. Laland, and Marcus W. Feldman are trying to change this:

> Organisms, through their metabolisms, activities and choices, define and partly create their own niches. They may also partly destroy them. This process of organism-driven environmental modification is called "niche construction." Niche construction regularly modifies both biotic and abiotic sources of natural selection and, in doing so, generates forms of feedback that change the dynamics of the evolutionary process.[28]

Obvious examples of niche construction are coral and the reefs that they build; beavers and their dams; and yours truly, *Homo sapiens,* and Paris.

Both of these biological mechanisms seem to involve a type of feedback that can alter the evolutionary process. The big idea behind the Baldwin effect is that sometimes both the direction and the rate of evolutionary change by natural selection can be affected by learned behaviors.

When one considers what has been happening in the last twelve thousand years, what we see is not a stable environment but a changing one, one where flexibility would be survival-enhancing. Not only was the landscape changing as the glaciers were retreating, but lifestyle, population density, and social organization were also changing. The question that presents itself is whether the increasing social interactions could in some way have affected our evolution. David Papineau makes an interesting point:

> [I]t has always seemed to me obvious that there is at least one kind of case where it [Baldwin effect] operates—namely, with the *social learning* of complex behavioural traits. . . . Suppose some complex behavioural trait P is socially learnt—individuals learn P from others, where they have no real chance of figuring it out for themselves. This will then create selection pressures for genes that make individuals *better* at socially acquiring P. But these genes wouldn't have any selective advantage without the prior culture of P, since that culture is in practice necessary for any individual to learn P. After all, there will not be any advantage to a gene that makes you better at learning P from others, if there aren't any others to learn P from. So this then looks like a Baldwin effect: genes for P are selected precisely because P was previously acquired via social learning. . . . Social learning has a special connection with the Baldwin effect because it is prone to trigger *both* of these mechanisms [genetic assimilation and niche construction]. When we have social learning, then we are likely to find cases where niche construction and genetic assimilation push in the same direction, and thus create powerful biological pressures.

The idea is that once individuals banded into groups, they were caught in a social world. Those who responded better to the social rules and practices that emerged were the ones who were more successful, survived, and reproduced. They were selected for, in a downwardly causal way, by the environment, part of which was social.

EVEN MONKEYS HAVE FUZZ

Complex social systems exist in other species, and clues to how ours arose are being teased from observations of these other animals. For instance, Jessica Flack has found evidence for the existence of monkey cops![29] These policing individuals are important to the cohesiveness of the social group as a whole. They not only terminate conflicts or reduce their intensity, but their presence also prevents conflicts from occurring and spreading, and it facilitates active sociopositive interactions among group members. When the policing macaques are temporarily removed, conflict increases. Just as with human societies, when the policeman is present, there are fewer barroom brawls, and speeders slow down on the freeways. Her results suggest that having a policeman around "influences large-scale social organization and facilitates levels of social cohesion and integration that might otherwise be impossible."[30] A macaque social network is more than just the sum of its parts. A group of macaques can foster either a harmonious, productive society or a divided, insecure grouping of cliques, depending on the organization of its individuals.

What is just as interesting, especially to our quest, is the conclusion she draws,

This means that power structure, by making effective conflict management possible, influences social network structure and therefore feeds back down to the individual level *to constrain individual be-*

haviour [italics added]. Pigtailed macaque social organization is not an epiphenomenon but a causal structure that both shapes, and is shaped by, individual interactions.

The social group constrains individual behavior, and individual behavior shapes the type of social group that evolves. This plays back to our idea of individual behavior's not being solely the product of an isolated, deterministic brain but being affected by the social group.

DOMESTICATING THE WILD MAN

Brian Hare and Michael Tomasello have proposed that the constraining of individual behavior has eventually led to genetic changes, as suggested by their emotional reactivity hypothesis. Chimpanzees are not generally a cooperative animal. They only cooperate in certain competitive situations—and only with certain individuals. This stands in marked contrast with humans, who are largely cooperative. Otherwise, how would the pyramids or the Roman aqueducts ever have been built? Hare and Tomasello think that the social behavior of chimps is constrained by their temperament, and the human temperament is necessary for more complex forms of social cognition. In order to develop the level of cooperation that is necessary for humans to live in large social groups, humans had to become less aggressive and less competitive. Hare and Tomasello think that humans may have undergone a self-domestication process in which overly aggressive or despotic others were either ostracized or killed by the group. Thus, the gene pool was modified, which resulted in the selection of systems that controlled (that is, inhibited) emotional reactivity such as aggression. (We will see later that an area of the right prefrontal cortex has actually been found that inhibits self-interested behavior!) The social group constrained the behavior and eventually affected the genome.

Hare's and Tomasello's emotional reactivity hypothesis grew out of

work done by Russian geneticist Dmitry Belyaev, who began domesticating foxes in Siberia in 1959 and whose domestication program continues today. He used only one criterion for his breeding selection
process: He picked the young foxes that approached his outstretched
hand the closest. Thus, he was selecting for fearless and nonaggressive
behavior toward humans. After only a few years, the by-products of this
selection process were similar to what is seen in domestic dogs. The
foxes had floppy ears, upturned tails, piebald colorations like border
collies, a prolonged reproductive season, and bigger litters; the females
had higher serotonin levels (known to decrease some types of aggressive behavior); and the levels of many of the chemicals in the brain that
regulate stress and aggressive behavior had been altered.[31] These domesticated foxes responded with equal skill to the human communica-
tive gestures of pointing and gazing as do domestic dogs.[32] All these
characteristics were linked to the gene associated with fear inhibition.
It seems that sociocognitive evolution has occurred in the experimental foxes as a correlated by-product of selection on systems mediating
fear and aggression. Dog domestication is thought to have occurred by
a similar process. Wild dogs that were less fearful of humans were the
ones that approached their camps, scavenged food, stuck around, and
reproduced. Perhaps both man's best human and canine friends were
selected for in the same manner.

SOCIAL TO THE CORE

The great social psychologist Floyd Henry Allport said "Socialized
behavior is . . . the supreme achievement of the cortex."[33] He was
right. If you think about this for a moment, you will realize that the
social world is our main focus, and it takes up an extraordinary
amount of our time and energy. When was the last time that you were
not thinking of something social? It shouldn't come as any surprise to
you that most of your thinking is social: Why are they doing that?

What was she thinking? Not another meeting! When did they get married? Does he like me? I owe them a dinner. And on and on. It can drive you crazy! All these social thoughts are reflected in our conversations. Consider all those cell phone conversations that you overhear. Ever hear anyone talking about particle physics or prehistoric stone axes? Social psychologist Nicholas Emler has studied the content of conversations and found that 80 to 90 percent are about specific names and known individuals, that is, social small talk.[34] We are social animals to the core.

THEORY OF MIND OR I KNOW THAT YOU KNOW THAT I BELIEVE THAT . . .

We neuroscientists are finally directing some of our efforts to the social world and the new field of social neuroscience has come on the scene. Complex social interactions depend on our ability to understand the mental states of others, and in 1978 David Premack came up with a fundamental idea that now governs so much of social psychological neuroscience work. He realized that humans have the innate ability to understand that others have minds with different desires, intentions, beliefs, and mental states, and the ability to form theories, with some degree of accuracy, about what those desires, intentions, beliefs, and mental states are. He called this ability theory of mind (TOM) and wondered to what extent other animals possessed it. Just the fact that he wondered if other animals possessed it sets him apart from most of us. Most people assume that other animals, especially cute ones with big eyes, have a theory of mind, and many of us even project this onto objects. In fact, within seconds, this response can be elicited in the presence of Leonardo, a socially programmed robot at MIT, who looks like a puckish cross between a Yorkshire terrier and a squirrel that is two and a half feet tall. While observing the behavior of what appears to be a self-propelled and goal-oriented robot, just as babies watching the triangle trying to get up the hill, we automatically see the robot as having intentions and we come up with psychological theories, that is, interpretations, about why Leo is behaving in a certain way, just as we do with other people (and our pets).

Once you understand the power of this mechanism, what activates it, and how we humans apply it to everything from our pets to our cars, it is easy to understand why anthropomorphism is so easy to resort to, and why it can be so hard for humans to accept that some of their psychological processes are unique. We are wired to think otherwise. After thirty years of clever research looking for TOM in other animals, evidence for it is lacking. It appears to be present to a limited degree in chimpanzees,[35] but that is it so far. So even though you have a theory about your dog, what he is thinking about, what he believes, and so forth, he does not have a theory about you, and he gets along quite well by tracking observables—your movements, facial expressions, habits of behavior, and tone of voice, and making predictions from them. TOM is fully developed automatically in children by about age four to five, and there are signs that it is partially,[36] or even fully, present by eighteen months.[37] Interestingly, children and adults with autism have deficits in theory of mind and are impaired in their ability to reason about the mental states of others,[38] and, as a result, their social skills are compromised.

MIRROR NEURONS AND UNDERSTANDING MENTAL STATES

In the mid-1990s, while they were studying the grasping neurons in macaque monkeys, Giacomo Rizzolatti and his colleagues discovered something quite remarkable and soon realized that they had come across the cortical origins of how an animal could appreciate the mental state of another. They found that when a monkey grasps a grape, the very same neuron fires as when the monkey observes another individual grasping a grape.[39] They called these mirror neurons, and they are one of the great recent discoveries in neuroscience. They were the first concrete evidence that there is a neural link between observation and imitation of an action, a cortical substrate for understanding and appreciating the actions of others. Since these original observations, mirror neuron systems that are quite different and much more extensive than those of the macaque have been identified in humans. The mirror neurons in the monkey are restricted to hand and mouth movements and only fire when there is goal-directed action, which may be why monkeys have very limited imitation abilities. In humans, however, there are mirror neurons that correspond to movements all over the body, and they fire even when there is no goal;[40] in fact, the same neurons are active even when we only imagine an action. The mirror neurons are implicated not only in the imitating of actions, but also in understanding the intention of actions.

UNDERSTANDING OTHER'S EMOTIONS

The ramifications of the human mirroring systems are gradually being understood and have huge implications. They are thought to be the neural basis of not only action understanding, but emotional understanding as well. In the insulae, humans have mirror systems, which are involved with understanding and experiencing the emotions of others, mediated through the visceromotor response.* Such systems, by unconsciously, internally replicating actions and emotions, may be the mechanism that gives us an implicit grasp of how and what other people feel or do, and contribute input used by the interpreter for theorizing about the cause (the why) for the actions and emotions of others. This is

* The response of the part of the motor system that controls the involuntary activity of smooth muscle fibers, the heart muscles, and glands (which secrete hormones).

known as simulation theory: You perceive through your senses an emotional stimulus (for example, you see the look of fear on someone's face), your body automatically responds to it by simulating it (you automatically imitate the look of fear, which results in your visceromotor system giving you a shot of adrenaline, thus simulating the emotion), which can either make it to your attention and be recognized or not. If it does come to your attention, then your interpreter comes up with a cause for the emotional feeling. You see your friend answer her phone and a look of happiness comes on her face. You smile too as you mirror her expression and you too get the same visceromotor response. You don't need to hear the other side of the conversation to know what your friend is feeling. You already know. You come up with the conclusion that she just got offered the job she was hoping for. We come to understand the states of others by simulating them in our brain and body.

These types of mirrored reactions have been demonstrated by fMRI scanning. For instance, there are anatomical connections in the brain between regions that make up the pain system that are highly interactive. There appears to be a separation, however, between the sensory (the ouch) and emotional (the anxiety of the "oh no, it's going to hurt") perceptions of pain. fMRI scans have shown that both the observer and the recipient of pain have activity in the part of the brain that is active with the emotional perception of pain, but only the recipient has activity in the area that is active with the sensory experience.[41] When you see another person in pain, you feel the anxiety, but not the pain itself. In another imaging experiment, subjects were first scanned while experiencing pain (either hot or cold stimuli) of different magnitudes to see what brain areas would be involved. The activity of one of the pain regions modulated according to subjects' reactivity to their pain: more pain, more activity. Then they were simply shown photographs of people experiencing pain (like a stubbed toe), and they rated the intensity of the pain they thought was being suffered. The same areas were active to the same degree both when they felt pain and when they looked at a painful image that they had rated of equal magnitude.[42] Taken together, these experiments are supportive of the idea that in order to understand the mental states of others, we literally simulate their mental state.

NONCONSCIOUS IMITATION OR MIMICRY

Our faces are our most prominent social feature. They reflect our emotional states, but as we have just learned, they also react to the emotional states of others. A 30-millisecond (ms) exposure to happy, neutral, and angry faces (too fast to consciously register that a face was even seen) will cause you to have measurable facial muscle reactions that correspond to the happy and angry faces[43] (these studies were done in nonsocial situations, which, we later see, matters). What we are talking about is nonconscious imitation, or mimicry. We actually mimic others constantly, but it happens so fast, we cannot actually perceive it.[44] We mimic the facial expressions, postures, vocal intonations, accents,[45] even speech patterns and words of others unconsciously.[46] Not only do we unconsciously copy the mannerisms of others, but we like and have smoother interactions with strangers if the stranger copies our mannerisms. Unconsciously, a connection is formed, and you "like" people who are similar to you. If we have been mimicked, we are also more helpful toward other people who are present than are nonmimicked individuals.[47] We also tend to agree with others we like.[48] Mimicry is what makes babies copy their mothers' expressions, sticking out their tongues when they do and smiling when they do. The consequences of this tendency to automatically mimic facial expressions, vocalizations, postures, and movements with those of another person are to converge emotionally with them, known as emotional contagion.[49] When one crying newborn in the nursery sets the others off, they are already showing evidence of emotional contagion.

Obviously all this mimicking behavior greases the machinery of social interactions and increases positive social behavior. This binding of people together through enhancing prosocial behavior may have adaptive value by acting as social glue that holds the group together, fostering safety in numbers.

When competition or members of a different group enter the picture, however, things change. People do not mimic the faces of those with whom they are in competition[50] nor with politicians with whom they do not agree.[51] More recently it has been shown that the relationship between the observer and the observed is relevant for mimicry reactions and not all emotional expressions are equally mimicked.[52] Happiness is always mimicked, negative expressions are not, depending on who is being mimicked. While mimicry increases rapport, it is not always beneficial to an individual to do so, especially with a competitor for limited resources. So while happiness, a low-cost emotion, is always mimicked, for it does not cost the observer, negative emotional expressions are mimicked only when shown by an ingroup member, because mimicking sadness (offering help) or anger (either signaling threat or expressing affiliation) can be costly. In fact, men only report sadness with a double affiliation: a close, ingroup member.[53] It is looking like mimicry is not purely automatic and reflexive; occasionally brakes are applied based on social context. It is an affiliative signal that is a major player in maintaining and regulating social interactions, especially within a social ingroup.

Voluntary imitation, however, is another cup of tea. It is difficult to consciously mimic someone because conscious behavior is slow. If you consciously try to mimic someone, it looks phony and throws the communication out of sync. Nevertheless, it is a tremendous way that things get transmitted among our species and is a potent mechanism in learning and acculturation.[54] Humans are the biggest voluntary imitators in the animal kingdom. We are, in fact, overimitators. While chimpanzees also voluntarily imitate others, they go straight for the goal or reward, while children will copy unnecessary actions to get a reward. Chimpanzees may imitate you as you cross over a plank for a banana, but they won't imitate your tiptoeing across as a child would do. Children are imitating machines, which is why parents have to be so careful about what they do and say, otherwise that cute lil' pumpkin is going to be swearing like a sailor. The ubiquitousness of imitation in the human world stands in stark contrast with its rarity in the animal kingdom. It appears to exist to some degree in the great apes, some birds, and perhaps in dolphins.[55] Even in all the thousands of monkeys that have been studied, voluntary imitation[56] has only been elicited in two Japanese monkeys after many years of intense training.[57]

INNATELY MORAL

We mirror, we imitate, we simulate emotions. We communicate in so many ways to navigate the social complexities of our human world. Even so, how is it that most of us get along—that 6.7 billion people aren't at one another's throats all the time? Are we really relying on learned behavior and conscious reasoning, or do we have a hardwired sense of appropriate behavior? Could we have an innate moral sense as a species that evolved as we banded together for survival? Is it not a good idea to kill because we are wired to think so, or because God or Allah or Buddha or our government said not to? These questions about whether we have an innate sense of moral behavior are not new. David Hume asked the same thing in 1777, "There has been a controversy started of late . . . concerning the general foundation of morals; whether they be derived from reason, or from sentiment; whether we attain the knowledge of them by a chain of argument and induction, or by an immediate feeling and finer internal sense."[58] While philosophers and religious leaders have been battling over this question for centuries, neuroscience now has tools and empirical evidence to help us answer it.

Anthropologist Donald Brown[59] collected a list of human universals that included many concepts that cultures share having to do with what is considered moral behavior. Some of these are: fairness;

empathy; the difference between right and wrong and redressing wrongs; praise and admiration for generous behavior; prohibitions against murder, incest, rape, and violent behavior; rights and obligations; and shame. Psychologist Jonathan Haidt, in an effort to include what is common to all moral systems, not just Western thought, has come up with this definition: "Moral systems are interlocking sets of values, virtues, norms, practices, identities, institutions, technologies, and evolved psychological mechanisms that work together to suppress or regulate selfishness and make social life possible."[60]

Moral Intuitions

Many moral intuitions are rapid automatic judgments of behavior associated with strong feelings of rightness or appropriateness. They are not usually arrived at by a deliberate conscious evaluative process that has been influenced by reason in the fullness of time. If you were to witness a person intentionally violate one of the above universal moral behaviors, most likely you would have a moral intuition about that behavior. A blatant example of such an intuition would occur if you were to see a child, who was quietly playing in a sandbox, get slapped in the face by his grandmother. You would have an immediate judgment about that behavior: It was bad, wrong, inappropriate, and you would be righteously indignant. When asked about your judgment it would be easily explicable. Such an example, however, doesn't really help us much when we consider Hume's question. Haidt came up with a different scenario and set about presenting it to all sorts of people:

> Julie and Mark are brother and sister. They are traveling together in France on summer vacation from college. One night they are staying alone in a cabin near the beach. They decide that it would be interesting and fun if they tried making love. At the very least it would be a new experience for each of them. Julie was already taking birth control pills, but Mark uses a condom too, just to be safe.

They both enjoy making love, but they decide not to do it again. They keep that night as a special secret, which makes them feel even closer to each other.[61]

Was it okay for them to make love? Haidt did a good job designing this story to stir up all of one's gut instincts and moral intuitions. He defines moral intuitions as "the sudden appearance in consciousness, or at the fringe of consciousness, of an evaluative feeling (like-dislike, good-bad) about the character or actions of a person, without any conscious awareness of having gone through steps of search, weighing evidence, or inferring a conclusion."[62] In his scenario, however, he also provides a rational answer to every objection. Haidt knew that most people would say that it was wrong and disgusting, and they do, but he wanted to get to the root reasoning, if any, we all must use. Why is it wrong? What does your rational brain say? Not unexpectedly, many answered that inbreeding could cause a deformed infant or that they could be hurt emotionally. Both of these objections, however, had been addressed in the original scenario. Haidt found that most respondents will eventually say, "I don't know, I can't explain it, I just know it's wrong." Is this a rational judgment or an intuitive one? Either we learned the moral rule that incest is wrong from our parents, religion, or culture or it is an innate, hardwired rule we have a difficult time over-ruling with rational arguments.

All cultures have incest taboos. It is universally accepted as bad human behavior. Because humans cannot automatically recognize their siblings by visual cues (hence all those movies where a sister and brother are raised separately, meet accidentally, and fall in love), Finnish anthropologist Edward Westermarck, in 1891, suggested that humans have evolved an innate mechanism, which usually works, that discourages incest. This mechanism causes a person to be uninterested or averse to having sex with people that one has spent a lot of time with as a child.[63] This rule predicts childhood friends and stepsiblings who were brought up together, along with full siblings, would all be found not to marry, and this has held true in studies looking at this question.[64]

Evolutionary psychologist Debra Lieberman expanded upon these findings.[65] She was interested how personal incest taboos—"Sex with my sibling is wrong"—generalize to a greater opposition, "incest is wrong for everyone," and whether this comes spontaneously from within or is learned. She found that an individual's moral attitude against incest in general was increased by the amount of time that the individual had actually spent living under the same roof with their sibling (related, adopted, or step-), and it was not increased by *learned* social or parental instruction, nor was it increased by the degree of relatedness to the sibling.

Shunning incest is not a rationally learned behavior and attitude that was taught to us by our parents or friends or religious teacher. If it were rational, then it would not apply to adopted or stepsiblings. It is a trait that has been selected by evolution because it worked in most situations to avoid producing offspring who were less healthy due to inbreeding and the expression of recessive genes. It is innate, and that is why it is universal in all cultures.

Your conscious rational brain, however, does not know that you have an innate incest avoidance system. All it knows is that siblings are having sex and that is BAD. It is when you are asked "why is it bad?" that your interpreter, working only with the information it has, which usually doesn't include the latest literature on incest avoidance but does include that bad feeling, tries to explain, and a variety of reasons will come pouring out of your brain!

That Ol' Trolley Problem

Another approach to the question of whether there is universal moral reasoning took Marc Hauser and his colleagues to the Internet with the classic trolley problem, devised by the philosophers Philippa Foot and Judith Jarvis Thomson. He predicted that if moral judgments are the result of a rational process, then people of different ages and cultures would have different answers for abstract moral problems. What is your answer?

> A runaway trolley is headed for five people, who will be killed if it proceeds on its present course. The only way to save them is for Denise, a passenger on the train, to pull a switch that will turn the trolley onto an alternate set of tracks, where it will kill one person instead of five. Should Denise pull the switch and turn the trolley in order to save five people at the expense of one?

Of the more than 200,000 people across the world who replied, 89 percent agreed that it is okay for Denise to pull the switch. But when asked this question:

> As before, a trolley threatens to kill five people. Frank is standing next to a large stranger on a footbridge crossing above the tracks, between the oncoming trolley and five workmen on the tracks below. If he pushes the large stranger off the bridge and onto the tracks below it will stop the trolley. The stranger will die if he does this, but the five workmen will not be killed. Is it okay for Frank to save the five others by pushing this stranger to his death?

Eighty-nine percent of people will answer NO to this one. This is a striking agreement across age and cultural groups, as is the dichotomy in response, when the actual numbers (save five by allowing one death) are no different in the two dilemmas. When people are asked about their response, whatever it is, they offer a wide variety of explanations, none particularly logical. Knowing what we have learned about our interpreter module, we would expect that there would be all sorts of explanations. The neuroscientist doesn't really care what they all are but wonders if there are moral reasoning centers or systems in the brain, what kinds of dilemmas activate them and what areas of the brain are active when moral decisions are being made.

Joshua Greene and his colleagues wondered if the same part of the brain was being used in both scenarios, so they scanned subjects in a brain imaging experiment while they were deciding their responses. With the first dilemma, which was an impersonal one (flip a switch), areas associated with abstract reasoning and problem solving had increased activity, and in the second scenario, which was a personal dilemma (the stranger had to be physically touched and pushed), the brain areas associated with emotion and social cognition had increased activity.[66] There are two different interpretations of these results. I have given a hint as to what Greene thinks is the difference: impersonal versus personal. Marc Hauser wasn't convinced, pointing out that there are too many variables in these dilemmas to narrow it down to personal versus impersonal. The responses could also be explained from the standpoint that the means don't justify the ends: the philosophical principle that it is permissible to cause harm as a by-product of achieving greater good but not to use harm to achieve it.[67] This is then discussing action based on intent. With either interpretation, the notion is that in certain situations there are moral brakes that are universal, and they come on and stop us from certain activities.

Moral Judgment and Emotions

Antonio Damasio and his group were able to help answer the question of whether emotional responses played a causal role in moral judgments.[68] They had a group of patients who had lesions in a brain region necessary for the normal generation of emotions, the ventromedial prefrontal cortex (VMPC). These patients had defects both in emotional response and emotion regulation, but normal general intelligence, logical reasoning, and declarative knowledge of social and moral norms. The Damasio team hypothesized that, if emotional responses (mediated by

VMPC) influence moral judgment, then these patients would make utilitarian judgments on the personal moral scenarios (the second trolley problem), but would have a normal pattern of judgment on the impersonal moral scenarios. While being scanned, the patients answered questions about situations with competing low-conflict-solution choices, such as, "Is it okay to murder your boss?" Both normal controls and people with the lesions answered, "No, that's crazy, it is not okay." Things changed, however, when the question was about high-conflict, personal moral dilemmas (which had competing considerations of aggregate welfare versus harm to others) that normally evoke a strong social emotion. Along with the second trolley problem, another such question would be: "During a brutal war, you are hiding from enemy soldiers in a room with ten other people including a baby. The baby starts to cry, which will reveal your hiding place. Is it okay to smother the baby so that the nine other people won't be discovered and killed by the soldiers?" With this type of question, the VMPC patients' judgment and reaction times significantly differed from the controls. Without the emotional reaction to the scenarios, they were faster and more utilitarian in their responses: Sure, push the fat man, and sure, smother the baby.

Moral Emotions, Moral Rationalizations, and the Interpreter

Jonathan Haidt proposes that people begin with their reaction to the dilemma, a result of an unconscious moral emotion, and then work backward to justify it. Here the interpreter steps in and comes up with a moral rationalization using information from the individual's culture, family, learning, and so forth. Although it is possible, we don't generally engage in moral *reasoning*. This only tends to happen when we change our perspective, put ourselves in the shoes of another person, see where they're comin' from. Marc Hauser suggests that we are born with abstract moral rules and a preparedness to acquire others, just as we are born with a preparedness to acquire language, and then the

environment, our family, and culture constrain and guide us to a particular moral system, as they do to a particular language.

Yet consider Steven Pinker's trolley scenario:

> A runaway trolley is about to kill a schoolteacher. You can divert the
> trolley onto a sidetrack, but the trolley would trip a switch sending a
> signal to a class of six-year-olds, giving them permission to name a
> teddy bear Muhammad. Is it permissible to pull the lever?
>
> This is no joke. Last month a British woman teaching in a private
> school in Sudan allowed her class to name a teddy bear after the
> most popular boy in the class, who bore the name of the founder of
> Islam. She was jailed for blasphemy and threatened with a public
> flogging, while a mob outside the prison demanded her death. To
> the protesters, the woman's life clearly had less value than maximiz-
> ing the dignity of their religion, and their judgment on whether it is
> right to divert the hypothetical trolley would have differed from
> ours. Whatever grammar guides people's moral judgments can't be
> all *that* universal. Anyone who stayed awake through Anthropology
> 101 can offer many other examples.[69]

While Pinker's objection presents a problem, it is not impossible to
square this up with our theory of a universal innate moral behavior; we
just have to consider the influence of culture, and Jonathan Haidt and
his colleague will help.

UNIVERSAL MORAL MODULES

Haidt and Craig Joseph have come up with a list of universal moral
modules after comparing works about human universals, cultural differences in morality, and precursors of morality in chimpanzees. Their
five modules have to do with suffering (it's good to help and not harm
others), reciprocity (from this comes a sense of fairness), hierarchy

(respect for elders and those in legitimate authority), coalitionary bonding (loyalty to your group) and purity (praising cleanliness and shunning contamination and carnal behavior).[70] Intuitive moral judgments arise from these modules, which evolved to deal with the specific circumstances common to our hunter-gatherer ancestors, who lived in a social world made up of groups of mostly related people, banded together for survival. Occasionally they came across other bands, some hostile, some not, some more closely related than others, all dealing with the same survival problems: limited resources, eating and not being eaten, finding shelter, reproducing, and taking care of offspring. They often faced dilemmas in their interactions, and some of these circumstances involved what we now consider to be moral or ethical issues. Individual survival was dependent on both the survival of the group that offered the protection of numbers, and his personal skills within the social group and with the physical world. Individuals and groups that survived and reproduced were those who navigated such moral issues successfully. Darwin recognized this when he wrote,

> A tribe including many members who, from possessing in a high degree the spirit of patriotism [coalitions], fidelity [coalitions], obedience [respect for authority], courage, and sympathy [suffering], were always ready to aid one another [reciprocity], and to sacrifice themselves for the common good [coalitions], would be victorious over most other tribes; and this would be natural selection. At all times throughout the world tribes have supplanted other tribes; and as morality is one important element in their success, the standard of morality and the number of well-endowed men will thus everywhere tend to rise and increase.[71]

Virtues Are Not Universal

Haidt's and Joseph's list of moral modules, and hence what they consider to be the moral foundations of different societies, is broader than that of other Western psychologists. They attribute this to the influence of not only Western culture, but the culture of politically liberal universities from whence these researchers arise. They suggest that the first two modules, which are focused on an individual, are what Western culture and liberal ideology are founded on, while the other three modules, which are focused on the survival of the group, are also incorporated in the morality of conservatives and other world cultures.

While the moral modules are universal, virtues, which are based on a hodgepodge from these modules, are not. Virtues are what a specific society or culture values as morally good behavior that can be learned. Different cultures place different values on various aspects of Haidt's five modules. The family, the social milieu, and the culture that we find ourselves in influence individual thought and behavior. Thus, what one culture, one political party, indeed even what one family considers to be virtuous (morally praiseworthy) is not universal. This is what drives cultural differences in morality and can explain Pinker's trolley problem. Haidt speculates that the differences between American political parties are driven by differences in value that they place on the five moral modules.

BELIEF ATTRIBUTION IN THE RIGHT BRAIN?

Neuroscientist Rebecca Saxe thought there was more than just the simulating of emotions going on when we tried to understand the beliefs and moral stance of others, or when we try to predict and manipulate their beliefs. To see if she was correct, she and her colleagues employed the classic false belief task while brain scanning her subjects. In the false belief task, Sally and Anne are in a room and Sally hides a ball in a blue box while Anne watches, and then Sally leaves the room. Anne then gets up and moves the ball into a red box. Then Sally comes back into the room. After watching this scenario and then being asked where Sally thinks the ball is, the children who are younger than four say Sally thinks the ball is in the red box. They do not understand that Sally has a false belief. After age four to five, they do understand and say that she thinks the ball in the blue box. This is a mechanism that develops and predictably comes on line at age four to five that allows understanding that others can have a false belief. Saxe has found that a specific right-hemisphere brain area is activated when adult subjects think about the beliefs of others, when they are explicitly told someone's belief in writing, when they follow loose directions to consider another's belief, and when they are instructed to predict actions of someone holding a false belief.

When I first heard of these findings, I was shocked that this mechanism was located in the right hemisphere. Because if this information about the beliefs of others is housed in the right hemisphere, and if, in split-brain patients, the information about others can't get to the left hemisphere, which does problem solving and has language capacity, then they should have a disruption in moral reasoning. But this doesn't happen. Split-brain patients act like everyone else. Once again my colleagues and I tested our endlessly patient patients. We took this idea that belief attribution of others is located in the right hemisphere, already knowing that goal representation of others is located in the left hemisphere, and we asked the following kinds of questions to our split-brain patients:

1. If Susie the secretary believes that she is adding sugar to her boss's coffee, but it actually was poison accidentally left there by a chemist, her boss drinks it and dies, was that a permissible action?

2. If Susie the secretary wants to bump her boss off and intends to add poison to his coffee, but it actually is sugar, he drinks it, and he is fine, was that permissible?

Is a listener to these stories going to be concerned about only the *outcome*, or will they judge on the basis of the *beliefs of the actor*? If you or I were asked these questions, we would say that the first action was permissible because she *thought* that the coffee was okay. The action in the second question, however, was not permissible, because the secretary *thought* it was poisoned coffee. We judge based on the secretary's intent, the beliefs of the actor. How were our split-brain patients going to respond? Separating the brain areas involved with the belief of others from the areas responsible for problem solving, language, and speech would predict that the split-brain patient would care only about outcomes, and this is what happened. They judged completely based on outcomes.

For example, after hearing a scenario in which a waitress *knowingly* serves sesame seeds to somebody she believes is highly allergic to them, yet the outcome was positive because the person turned out not to be allergic, J.W. quickly *judged the action to be permissible*. Because the split-brain patients function normally in the real world, what happened next wasn't surprising. Moments later, after his conscious brain processed what he had just said, J.W. rationalized (the interpreter to the rescue) his response: "Sesame seeds are just little, tiny things, they are not going to hurt you." He had to square his automatic response, which did not benefit from information about the belief state of the waitress, with what he rationally and consciously knows is permissible in the world.

INHIBITING SELF-INTEREST

We often consider dilemmas that have to do with fairness to be moral dilemmas. A fascinating, well-known finding involves what is known as the ultimatum game. Two people are involved in this game and they are only allowed one round. One person is given twenty dollars, and he has to split it with the other player, but he determines the percent split.

Both players get to keep whatever amount of money is first offered. However, if the player who is offered the money refuses the offer, then neither gets any. In a rational world, the player who gets offered the money should take any offer because that is the only way he will come out ahead. That, however, is not how people react. They will accept the money only if they think it is a fair offer, ranging from at least six to eight dollars. Ernst Fehr[72] and his colleagues used transcranial electric stimulation to disrupt brain functioning in the prefrontal cortex and found that when the function of the right dorsolateral prefrontal cortex was disrupted, people would accept lower offers while still judging them to be unfair. Since suppression of this area increased selfish responses to unfair offers, it suggests that this area normally inhibits self-interest (taking any offer) and reduces the impact of the selfish urges on the decision-making processes, and thus plays a key role in implementing behaviors that are fair.

More evidence for this region's inhibiting selfish responses is from Damasio's group, which has given moral tests to adults who have had injuries to this area since childhood. Their answers were excessively egocentric, as was their behavior. They exhibited a lack of self-centered inhibition and did not take another's perspective. People who acquire these types of lesions as adults, such as the patients Damasio tested with the moral dilemma problems, can compensate better, which suggests the neural systems that had been impaired at an early age were critical for the acquisition of social knowledge.[73]

Many examples of moral circuits have been identified, and they seem to be distributed all over the brain. We have many innate responses to our social world, including automatic empathy, implicit evaluation of others, and emotional reactions, and these all inform our moral judgments. Yet we typically do not think about these automatic responses nor appeal to them when explaining our decisions. Humans act commonly on moral challenges but claim different reasons for doing so. This is because there is a cacophony of influences that guide our behavior and judgments. The influences involve emotional systems and special moral judgment systems; the innate moral behavior pours out, and then we give it an interpretation. We personally believe the interpretation and it becomes a meaningful part of our life. But what sets off our responses are these universal properties that we all have.

It appears we all share the same moral networks and systems, and tend to respond in similar ways to similar issues. The way we differ is not our behavior but our theories about why we respond the way we do and the weight that we give these different moral systems. Understanding that our theories and the value that we place on them are the source of all our conflicts would go a long way, it seems to me, in helping people with different beliefs systems to get along.

Our brain has evolved neural circuitry that enables us to thrive in a social context. Even as infants we make judgments and choices and behave based on the action of others. We prefer others who are helpful, or even neutral, to others who hinder. We understand when another needs help, and we engage in altruistic helping. Our extensive mirror neuron systems give us the ability to understand the intentions and emotions of others, and from this information our interpreter module weaves together a theory about others. We also use the same module to weave a story about ourselves.

As our social context changes through the accumulation of knowledge about our very nature, we may want to change how we live and experience our social life—especially with respect to justice and punishment. This leads us to the story in the next chapter about how we incorporate social dynamics into personal choice, how we figure out the intentions, emotions, and goals of others in order to survive, and understand how social process constrains individual minds.

Chapter Six

WE ARE THE LAW

On February 19, 1997, a house painter called 911 in Tampa, Florida. He had returned unannounced to a client's house and through a window saw what appeared to be a naked man throttling a naked woman. When the police arrived, a neighbor said that a man "came out of the house staggering. His shirt was unbuttoned, and he had blood all over his chest."[1] The man hadn't just throttled the woman, he had stabbed her multiple times, killing Roxanne Hayes, a mother of three children, aged three to eleven. His name was Lawrence Singleton; he was seventy years old, and he was notorious in California, where nineteen years before, he had raped a fifteen-year-old hitchhiker, Mary Vincent, hacked off her forearms with an ax, and left her in a culvert beside the road of Del Puerto Canyon to die. Two vacationers came across her the next morning, walking naked toward the interstate, the stumps of her severed arms raised to prevent further blood loss. Vincent's description of her attacker was so vivid that it resulted in a police artist's drawing that his neighbor recognized. Singleton was tried, found guilty, and given what was the maximum sentence at the time in California of fourteen years, but was released on parole after eight years of "good behavior," even though shortly before his release a prison psychiatric evaluation read, "Because he is so out of touch with his hostility and anger, he remains an elevated threat to others' safety inside and outside prison."[2] Mary's mother, Lucy Vincent, said that Mary's father would carry a .45-caliber pistol and often contemplated killing Singleton.[3] After his parole, Mary was terrified for two reasons. While in prison Singleton had written letters to her lawyer threatening her; and after she had testified and walked past him in the courtroom he whispered, "I'll finish this job if it takes me the rest of my life."[4] After his parole, she was afraid to stay in one place too long and had numerous bodyguards.

In 1997 Mary told a St. Petersburg Times reporter, "I'm not paranoid enough." It wasn't just Mary who was paranoid. After he was paroled, residents of every California town that prison authorities tried to settle him in staged angry protests. He was finally settled in a mobile home on the grounds of San Quentin prison until his parole was up. Outrage in California over Singleton's parole led to the Singleton bill, preventing the early release of offenders who have committed a crime involving torture, and changed the sentence for such crimes to a twenty-five-years-to-life sentence.[5] In 2001 Singleton died of cancer while on death row in Florida. Mary Vincent told a reporter that the arrest and death of the man responsible had given her a "tremendous feeling of freedom," but that she still has nightmares and is afraid to go to sleep. "I've broken bones thanks to my nightmares. I've jumped up and dislocated my shoulder, just trying to get out of bed. I've cracked ribs and smashed my nose."[6] Divorced, with prostheses that she has modified with spare parts from broken refrigerators and stereo systems, she is now an artist struggling to support two sons.

While you read this, what were your gut feelings and thoughts about Larry Singleton? Did you want him to be locked up and never released (incapacitation)? If you had been Mary's father, would you have wanted to kill him (retribution)? Or did you want to forgive him, to tell him that it is too bad his brain was unable to inhibit his naturally aggressive tendencies and perhaps with some treatment he could be more prosocial (rehabilitation)? Incapacitation, retribution, or rehabilitation are the three choices society has for dealing with criminal behavior. When society considers public safety, it is faced with the decision about which perspective those making and enforcing the laws should take: retribution, an approach focused on punishment of the individual and just deserts; or consequentialism, a utilitarian approach that what is right is what has the best consequences for society.

As neuroscience comes to an increasingly physicalist understanding of brain processing, it is beginning to challenge some people's notions about criminal behavior and what we should do about it. Determinism disputes long-standing beliefs about what it means to be responsible for one's actions, with some scholars asserting the extreme view that humans are never responsible for any of their actions. These ideas challenge the very foundational rules regulating how we live together in social groups. Should people be held accountable for their behavior? If they aren't, it seems that it would change behavior for the worse, just as reading about determinism results in increased cheating on tests, as we learned in chapter four and would adversely affect society in general. Is accountability what keeps us civilized? Neuroscience has more and more to say about these questions and is already slowly oozing into the courtroom—prematurely, to the view of most neuroscientists.

Californians thought that Singleton should not have been paroled, that he was still a threat, and they didn't want him in their communities. They also thought that certain behavior warranted longer incarceration. Unfortunately, in his case, they were right, and the parole board was wrong. More recently, the legal system has been looking to neuroscience to provide answers in several different arenas: predicting a person's future threat (recidivism), determining for whom treatment is possible, and deciding what level of certainty about these determinations is acceptable. Are some crimes just too horrendous to contemplate release? Neuroscience is also illuminating why we have the emotional reactions that we do to antisocial or criminal behavior. This leads us to the question that if we understand our reactions that have been honed by evolution, can or should we amend them? Are these emotions the sculptors of a civilized society? We have our work cut out for us!

The title of this chapter, "We Are the Law," was suggested to me by the philosopher Gary Watson, who pointed out the simple fact that as we come to think about ourselves, we shape the rules that we decide to live by. If Michael Tomasello and Brian Hare are correct that we have been domesticating ourselves over thousands of years through ostracizing and killing those who were too aggressive, in essence removing them from the gene pool and modifying our social environment, then we have been making rules for groups to live by and enforcing them throughout our evolutionary history. If, because of the findings of the various branches of neuroscientific research that I have been presenting to you, we come to think differently about ourselves, our behaviors, and motivations from what we had thought two or three hundred years ago, then we *may* decide to reconstruct our social framework. This comes down to the fact that we are the law because we make the laws. We have a stance balanced by innate views of moral thinking and culturally specific ideas. As we look into issues of how brain enables mind, we are asked to decide if we must come to a different belief about the nature of man, about what we are, and how we should interact. It may follow that inevitably we will consider whether it would be beneficial or not to change our legal structure.

So far we have seen that the mind constrains the brain, and we have come to understand that social process constrains individual minds. In this chapter we are going to see that the views emerging from neuroscience about the human condition are having a cultural impact on law and on our concepts of responsibility and justice. The questions being chewed over are at the very foundation of our legal system: Is our natural inclination for retribution necessary, or is utilitarian accountability sufficient? Is punishment justified? I won't keep you in suspense.

These are questions that haven't in any way been answered, but they are brought to the fore by research on the brain and what it tells us about who we are. We are going to see that our current legal system has emerged from innate intuitions, honed by evolution, just as our moral systems have been.

CULTURE AND GENES AFFECT COGNITION

The culture to which we belong actually plays a significant role in shaping some of our cognitive processes. This idea was looked into by Richard Nisbett and his colleagues, who posited that East Asians and Westerners actually use different cognitive processes in thinking about certain things, and that the origins of these differences were in their different social systems, one arising out of the civilization of ancient China and the other out of ancient Greece.[7] They characterize the ancient Greeks as having no counterpart among the other ancient civilizations, and remarkable in regards to their locating power within the individual. Nisbett, writing about his findings, states, "The Greeks, more than any other ancient peoples, and in fact more than most people on the planet today, had a remarkable sense of personal *agency*— the sense that they were in charge of their own lives and free to act as they chose. One definition of happiness for the Greeks was that it consisted of being able to exercise their powers in pursuit of excellence in a life free from constraints."[8] Ancient Chinese differed in that their focus was on social obligation or collective agency. "The Chinese counterpart to Greek agency was *harmony*. Every Chinese was first and foremost a member of a collective, or rather of several collectives—the clan, the village, and especially the family. The individual was not, as for the Greeks, an encapsulated unit who maintained a unique identity across social settings." With harmony as the goal, confrontation and debate were discouraged.

Nisbett and colleagues suggest that social organization affects cognitive processes indirectly by focusing our attention on different parts of the environment and directly by making some social communication patterns more acceptable than others. The idea is if one sees oneself as an interwoven part of a big picture, then one might look at all aspects of the world holistically, whereas if one sees oneself as having individual power, one looks at aspects of the world individually. And that is what has been demonstrated. In tests where Americans or East Asians described simple scenes that were flashed to them and later were tested on what they remembered from the scenes, Americans focused on the main items in a picture, whereas Asian viewers attend to the entire scene. Is this cultural difference evident in brain function?

It seems to be. At MIT, researchers Trey Hedden and John Gabrieli had East Asians and Americans make quick perceptual judgments while having an fMRI scan.[9] They were shown a series of differently sized squares and each square had a single line drawn inside of it. The Americans, judging whether the line-to-square-size proportion was the same or different from one square to the next (a relative judgment), showed much more brain activity was needed for sustained attention than when judging if lines were of the same length regardless of the surrounding squares (an absolute judgment of individual objects). For them, absolute judgments about individuals required less work by the brain, but judgments about relationships used more. The exact opposite was true for the East Asians. Their brains had to work overtime on the absolute judgments, but breezed through the relative ones. In addition, the amount of activity for the culturally preferred and nonpreferred tasks also varied according to the degree that the individual identified with his culture. The differences in brain function occurred during the late stage of processing when attention is focused on the judgment, not during the early stage of visual processing. While the same neural systems were used by both groups, they differed in magnitude for the type of task "completely reversing the relation between task and activation across a widespread brain network."

These different styles of focusing attention are also found within the same geographic region and ethnic group. The fishermen and farmers of the eastern Black Sea region of Turkey, who live in communities based on cooperation, tended more toward holistic attention than the shepherds, who live in communities where individual decisions are constantly being made.[10]

Easterners and Westerners also vary in their genetic makeup, and Heejung Kim and her colleagues wondered to what extent genetic differences can account for differences in attention. Much research had already shown that serotonin plays a part in attention, cognitive flexibility, and long-term memory, so they decided that looking into a specific serotonin system polymorphism (a DNA sequence variation), which was known to affect an individual's mode of thinking, could prove fruitful. They looked at different alleles (genes which have different nucleic acid sequences occupying the same position on a paired chromosome that control the same inherited characteristic) of the 5-HTR1A gene that ultimately controls the neurotransmission of serotonin. They found that there was a significant interaction between the type of 5-HTR1A alleles a person had and the culture in which he lived. This interaction affected where that particular person's attention was directed. Those persons possessing the identical DNA sequences in the matched gene pairs (homozygous) G allele, which is associated with the reduced ability to adapt to changes, more strongly endorse the culturally reinforced mode of thinking than those with the homozygous C allele. Those who possessed one G and one C sequence (heterozygous G/C allele) had a middle-of-the-road opinion! Summarizing their findings, theses researchers concluded, "The same genetic predisposition can result in divergent psychological outcomes, depending on an individual's cultural context."[11]

It is powerful to see that behavior, cognitive stance, and underlying physiology affect and can be affected by the cultural milieu. This strengthens the importance of the idea of the niche construction model that I described in the last chapter, where interactions between organisms and their environment are bidirectional: The organism (or the selectee) actually changes the environment (the selector) somewhat, perhaps affecting the results of future selection. For example, in the case of humans, we have the ability to change the environment, not only physically but socially, and the feedback from these changes produce a changed environment, which selects which humans will survive and reproduce and cause future changes to the environment. Thus the environment and the organism are coupled across time.

These ideas become particularly important when you consider how our legal structures and moral rules affect and shape our social environment, what behaviors they may be selecting for, who will survive and reproduce, and how that will affect future social environments. On the neurophysiological level, we are born with a sense of fairness and some other moral intuitions. These intuitions contribute to our moral judgments on the behavioral level, and, higher up the chain, our moral judgments contribute to the moral and legal laws we construct for our societies. These moral laws and legal laws on the societal scale provide feedback that constrains behavior. The social pressures on the individual at the behavior level affect his survival and reproduction and thus what underlying brain processes are selected for. Over time, these social pressures begin to shape who we are. Thus, it is easy to see that these moral systems become real and very important to understand.

WHO DONE IT, ME OR MY BRAIN?

Legal systems serve as a social mediator of dealings between people. We should keep in mind the niche construction dynamic when attempting to characterize the law and our concepts of justice and punishment, formed, as they were, by the human brain, mind, and cultural interactions. Legal systems elaborate rights and responsibilities in a variety of ways. In most modern day societies, the laws made by these systems are enforced through a set of institutions as are the consequences of breaking those laws. When one breaks a law, it is considered to be an offence against the entire society, the state, not an individual. Currently, American law holds one responsible for one's criminal actions unless one acted under severe duress (a gun pointed at your child's head for instance) or one suffers a serious defect in rationality (such as not being able to tell right from wrong). In the United States, the consequences for breaking those laws are based on a system of retributive justice, where a person is held accountable for his crime and is meted out punishment in the form of his "just deserts." After the previous chapters and the evidence for determinism, we are confronted with the question: Who do we blame in a crime, the person or the brain? Do we want to hold the person accountable or do we want to forgive him because of this determinist dimension of brain function? Ironically, this question is treading dualist waters, suggesting that there is a difference between a person and his brain and body.

NEUROSCIENCE OOZING INTO THE COURTROOM

The law is complicated and takes into consideration more than just the actual crime. For example, the intention of the perpetrator is also part of the equation. Was the act intentional or accidental? In 1963, Lee Harvey Oswald had the intention of killing President Kennedy when he took his concealed rifle to the building along the parade route, waited there until the president's motorcade was passing, and shot him. In an Australian case the following year, however, Robert Ryan was judged not to have had the intention to murder when he killed the cashier of a store he had just successfully robbed. While leaving the store, he tripped, accidentally pulled the trigger of his gun, and shot the cashier. While movies, books and television portray crimes ending up in a courtroom where intention and many other circumstances are examined, very few criminal cases actually go to trial, only about three percent; most are plea bargained out. Once we step into the court room, the laboratory of judicial proceedings, neuroscience has an enormous amount to say about the goings on. It can provide evidence that there is unconscious bias in the judge, jury, prosecutors and defense attorneys, tell us about the reliability of memory and perception with implications for eyewitness testimony, inform us about the reliability of lie detecting, and is now being asked to determine the presence of diminished responsibility in a defendant, predict future behavior, and determine who will respond to what type of treatment. It can even tell us about our motivations for punishment.

Robert Sapolsky, professor of neurology at Stanford, makes the extremely strong statement: "It's boggling that the legal system's gold standard for an insanity defense—M'Naghten—is based on 166-year-old science. Our growing knowledge about the brain makes notions of volition, culpability, and, ultimately, the very premise of a criminal justice system, deeply suspect."[12] The M'Naghten rules arose after the attempted assassination of the British Prime Minister Robert Peel in 1843 and have been used to determine (with a few adjustments) criminal liability in regards to the insanity defense in most common law jurisdictions ever since. The British Supreme Court of Judicature, in answer to one of the questions posed to it by the House of Lords about the insanity law, responded "the jurors ought to be told in all cases that every man is presumed to be sane, and to possess a sufficient degree of reason to be responsible for his crimes, until the contrary be proved to their satisfaction; and that to establish a defence on the ground of insanity, it must be clearly proved that, at the time of the committing of the act, the party accused was labouring under such a defect of reason, from disease of the mind, as not to know the nature and quality of the act he was doing; or, if he did know it, that he did not know he was doing what was wrong."[13] The question that Sapolsky raises is: Given determinism, given that we are beginning to understand mental states, given we can track down which part of the brain is involved in volitional activity and that it may be impaired, and our growing knowledge that we can be specific about the existence of an impairment and what is causing it, will we view the defendant differently?

At stake in the arguments is the very foundation of our legal system, which holds a person responsible and accountable for his actions. The question is this: Does modern neuroscience deepen our ideas about determinism, and, with more determinism, is there less reason for retribution and punishment? Put differently, with determinism there is no blame, and, with no blame, there should be no retribution and punishment. This is the simmering idea that people are worried about. If we change our mind about these things as a culture, then we are going to change how we deal with this unfortunate aspect of human behavior involving crime and punishment.

WOWED BY SCIENCE

Common Law is based on the belief that it is unfair to treat similar facts differently on different occasions, so "precedent" or past decisions bind those of the future. Thus, it is the past judgments of judges and juries that make common law, not legislative statutes. Looking back on the history of common law, its roots and many of its traditions were founded during a time when there wasn't much scientific knowledge available. Even as recently as the 1950s, what was admissible as science in the courtroom was psychoanalytic theory, which was not backed up with empirical data. Why was something with no empirical guts admissible? Because a judge had thought it was good enough and ruled it so. In the last half century things have changed. We have come a long way in our knowledge about brain function and behavior and do have empirical data. Now that we know all these brain mechanisms, the correlates of cognitive states and mental outlooks, brain scans have started to appear in the courtroom, admissible as evidence, to explain why someone acted in a particular way. Can these scans actually do this?

A majority of neuroscientists is not convinced that they can at this point, because when you read a brain scan, you are merely noting that in this particular area, if you average together several brains, such and such occurs in this location. A scan result is not specific for a specific person. This raises the question of why they are in the courtroom. It is hard not to think that there is something about our culture that actually believes more about scans than the scientist does himself. Yet attorneys and neuroscientists both wonder if these scans are more probative or are they prejudicial. Equally controversial is whether a judge or jury, untrained in science, can understand its limitations and the fallibility of interpretive conclusions. Many neuroscientists worry that a scientist who walks into a courtroom, shows a series of brain scans, and says this is why the defendant shouldn't be held responsible, is overly influential. Recent studies have shown that when adults read the explanations of psychological phenomena, the explanations are more positively evaluated and considered important if a brain scan is shown in the material they read, even when they have nothing to do with the explanations! In fact, bad explanations are more accepted with the presence of a brain scan.[14] This certainly seems to raise a red flag that jurors and judges could be primed by what is being presented as scientific certainty, when in reality, what scientists are reading in a brain scan it is a probabilistic calculation of where brain activity is taking place, based on averages of the activity in different individual's brains. We are going to get to this in a bit, but what is important to understand is one can't point to a particular spot on a brain scan and state with 100 percent accuracy that a certain thought or behavior arises from activity in that area. In games where students are to impose hypothetical punishments, if they have first read an excerpt about determinism (been primed for determinism), then they give less punishment than those who have not.[15] So what we come to believe about brain function is going to influence who we are and what we do.

The three areas of the law that neuroscience is now impacting have to do with responsibility, evidence, and the question of justice for the victim and the offender during sentencing.

Responsibility

In terms of responsibility, the law looks at the brain in this simple pattern: There is what is called a "practical reasoner" that is working freely in a normal brain producing action and behavior. Personal responsibility is a product of a normally functioning brain of the "practical reasoner." Things can happen to the brain, a lesion, injury, stroke, or neurotransmitter disorder that makes it not function normally, resulting in diminished brain capacity, thus, diminished responsibility, and this is used for exculpability. In criminal cases in particular, the defendant must also have "mens rea" or actual evil intent. One recent case in which brain scans were used to change two separate death sentences was in Pennsylvania. Simon Pirela had received two death sentences for two separate first degree murder convictions in 1983. In 2004, however, twenty-one years later, brain scans, allowed as evidence, convinced one jury in a resentencing hearing (that had been ordered due to prosecutorial misconduct) that Pirela was not eligible for the death penalty because he suffered from aberrations in his frontal lobes, diminishing his capacity to function normally. In an appeal to vacate the second death sentence, the same scans were used to make the different claim that Pirela was mentally retarded, which combined with the neuropsychologists' testimony were found "quite convincing" by the appellate judge.[16] The same scans were accepted as evidence for two different diagnoses.

It is interesting to note such cases are now being decided after the landmark case of *Atkins v. Virginia* (2002) in which the Supreme Court ruled that it was a violation of the Eighth Amendment of the U.S. Constitution to execute someone with mental retardation, as it would be cruel and unusual punishment. The Atkins case was summarized by Chief Justice Scalia as follows:

> After spending the day drinking alcohol and smoking marijuana, petitioner Daryl Renard Atkins and a partner in crime drove to a convenience store, intending to rob a customer. Their victim was Eric Nesbitt, an airman from Langley Air Force Base, whom they abducted, drove to a nearby automated teller machine, and forced to withdraw $200. They then drove him to a deserted area, ignoring his pleas to leave him unharmed. According to the co-conspirator, whose testimony the jury evidently credited, Atkins ordered Nesbitt out of the vehicle and, after he had taken only a few steps, shot him one, two, three, four, five, six, seven, eight times in the thorax, chest, abdomen, arms, and legs.
>
> The jury convicted Atkins of capital murder. At resentencing . . . the jury heard extensive evidence of petitioner's alleged mental retardation. A psychologist testified that petitioner was mildly mentally retarded with an IQ of 59, that he was a "slow learne[r]," . . . who showed a "lack of success in pretty much every domain of his life," . . . and that he had an "impaired" capacity to appreciate the criminality of his conduct and to conform his conduct to the law. . . . Petitioner's family members offered additional evidence in support of his mental retardation claim. . . . The State contested the evidence of retardation and presented testimony of a psychologist who found "absolutely no evidence other than the IQ score . . . indicating that [petitioner] was in the least bit mentally retarded" and concluded that petitioner was "of average intelligence, at least."
>
> The jury also heard testimony about petitioner's sixteen prior felony convictions for robbery, attempted robbery, abduction, use of a firearm, and maiming. . . . The victims of these offenses provided graphic depictions of petitioner's violent tendencies: He hit one over the head with a beer bottle . . . ; he slapped a gun across

another victim's face, clubbed her in the head with it, knocked her to the ground, and then helped her up, only to shoot her in the stomach, *id.* . . . . The jury sentenced petitioner to death. The Supreme Court of Virginia affirmed petitioner's sentence. . . .[17]

Writing for the majority of the Court, Justice Stevens reasoned that the two main justifications for capital punishment, deterrence and retribution, could not be appreciated by the defendant who suffered mental retardation, and therefore was cruel and unusual punishment. He did not address the third justification of capital punishment, which is incapacitation. In short the legal decision was delivered in terms of existing *beliefs* about the purpose of punishment in the law. It was not based on the science, namely that the defendant, because of his brain abnormality could or could not form intentions and the like. It also makes the supposition that anyone suffering any degree of "mental retardation" has no capacity for understanding the "just deserts" for a crime or what the society considers right or wrong.

There are other problems with the abnormal brain story, but the biggest one is that the law makes a false assumption. It does not follow that a person with an abnormal brain scan has abnormal behavior, nor is a person with an abnormal brain automatically incapable of responsible behavior. Responsibility is not located in the brain. The brain has no area or network for responsibility. As I said before, the way to think about responsibility is that it is an interaction between people, a social contract. Responsibility reflects a rule that emerges out of one or more agents interacting in a social context, and the hope that we share is that each person will follow certain rules. An abnormal brain does not mean that the person cannot follow rules. Note that in the above case, the perpetrators were able to make a plan, take with them what was necessary to implement the plan, understood that what they were doing was not something that should not be done in public, and were able to inhibit their actions until they were in a deserted area.

In the case of an abnormal neurotransmitter disorder such as schizophrenia, while there is a higher incidence of arrest for drug-related issues, there is no higher incidence of violent behavior in people with schizophrenia while they are taking their medication and only a very small increased incidence of those who are not. They still understand rules and obey them; for instance, they stop at traffic lights and pay cashiers. It is not true that just because you have schizophrenia your base rate of violent behavior goes up and you are vastly more likely to commit a crime. Using the defense of schizophrenia may help the defendant in one case, but it will also improperly liberate one in another case. It may also be used as evidence of a false accusation. Such thinking can also lead to the utilitarian practice of locking up all people with schizophrenia "before they commit a crime." Diagnosed with schizophrenia after the fact by a psychiatrist for his defense, John Hinckley was found not guilty by reason of insanity for his attempt to assassinate President Reagan. This attempt, however, was premeditated. He had planned it in advance, showing evidence of good executive functioning. He understood that it was against the law and concealed his weapon. He knew that shooting the president would give him notoriety. The same false assumption is also true for people who have acquired left frontal lobe lesions. They can act oddly: They, their family, and friends will notice changed behavior, but their violence rate only increases from the base rate of 3 percent to 11–13 percent. A frontal lobe lesion is not a predictor of violent behavior. There is no lesion in a specified location, no switch that turns on violent behavior. One case cannot generalize to another. If the court system concludes that frontal lobe lesions make a person exculpable for their behavior, then it may be left with people who have such lesions using their injuries as an excuse for things they wouldn't commit had they not this ready-made excuse (Great, I can knock off that jerk, and I'll just blame it on my frontal lobe and get off). Or, all people with frontal lobe lesions could be locked up as a prophylactic measure. So in thinking about these things, we have to be careful that our best intentions aren't used in an inappropriate way.

Evidence

How did psychoanalytic theory, and now brain scans, become admissible in court? In the United States, there are general standards for scientific evidence to be admissible in court. Various states follow either the Frye rule of general acceptance, which states "scientific evidence is admissible when the scientific technique, data, or method has 'gained general acceptance' by the relevant community,"[18] or the Daubert-Joiner-Kumho "validity" rule,* where trial judges possess "gatekeeping responsibility" in determining validity of scientific evidence and all expert testimony, or a combination of both. Judges use several criteria, such as whether a theory or technique is falsifiable or has been subjected to peer review and so forth, to analyze whether expert testimony is good science, but can a judge, trained in law, reliably judge if scientific evidence is valid?

Brain images, whether they should have been admissible by scientific standards or not, have made their way into the courtroom and we have to deal with them. Functional brain imaging is the basis for the growing tendency to think of the brain in deterministic terms, even though the newer scans are far more statistical in nature, as discussed below. Nonetheless, it seems inevitable that the findings of functional brain imaging examinations will also be introduced as evidence in legal proceedings. Closer inspection of the technique, however, should cast doubt on these interpretations and expectations.

ONE BRAIN FITS ALL? THE PROBLEM OF INDIVIDUAL VARIATION

Like fingerprints, everyone's brain is slightly different, has a unique configuration, and each of us reliably solves problems in different ways.

* Standards for Fed. Rule Evid. 702.

That is not news to anyone, and there is a rich history of individual variation in psychology. This fact was put aside for a while, however, when brain scanning was first being developed. Having a beautiful brain scan is one thing, knowing what you are looking at, what an area's function is, how it relates to other areas of the brain, how to localize a particular structure from one brain to the next were all unknowns. MRI scans vary greatly from individual to individual because of differences in brain size, shape, and differences in slice orientation due to these variations and also the programming of the scanner, and so forth. In 1988 Jean Talairach and Pierre Tournoux published a three-dimensional proportional grid system to identify and measure brains despite their variability. The system is based on the idea that brain components, deep within its structure that cannot be seen from its surface, can be defined in relation to "two easily identifiable features on the brain's surface, the anterior and posterior commissures." Using these standard anatomical landmarks, individual brain images obtained from MRI and PET scans can be morphed on the "standard Talairach space." Using their atlas, inferences can be drawn about tissue identity at a specific location.

There are limitations to this method and Talairach was quick to point out that the brain he used for reference (the postmortem brain of a sixty-year-old Frenchwoman) to construct the standard space was a smaller-than-average brain and "Because of the variability in brain size, specifically at the level of the telencephalon,* this method is valid *with precision* [italics added] only for the brain under consideration."[19] That is, he is saying that it is only precise for that particular sixty-year-old Frenchwoman's brain that was smaller than average. "Normalization software," which rotates, scales, and perhaps warps the brain to fit the standard template, is used to compare brains, starting out by first smoothing out the sulci (the deep grooves in the cerebral surface) in

---

* The anterior portion of the brain that is comprised of the cerebral cortex, the olfactory bulb, the basal ganglia, and the corpus striatum.

the brain images, which are widely variable between individuals. In doing so, it loses the detail of sulcal information and does not result in consistent sulcal locations. Thus, the coordinates of where a specific area is located are probabilistic, with variation in the actual location from one individual to the next. In turn, the location in the brain of any particular brain process is also probabilistic and not precise, but is the best that can be currently done without directly examining a brain. Neuroscience's own little Uncertainty Principle!

In order to establish a standard for the *workings* of the brain through imaging, the signal to noise ratio, that is the signal of interest amid all the other brain signals, had to be high enough to indicate that a particular response had occurred in a particular location. To do this, Michael Miller and his colleagues at Dartmouth College scanned the brains of twenty people, morphed all the separate brain scans into one, and added all the signals onto that averaged morphed brain. The regions where the signals were consistently present indicated that that area could be reliably identified as being active for that task across individuals. If most of the information about brain work comes from group averages like this, however, how do you get to the individual? How do you get to the defendant in the courtroom? For instance, if you look at the group map for a recognition memory task where you remember something seen previously, the average result of sixteen subjects shows that the left frontal areas are heavily involved in this type of memory task.[20] When you look at the individual maps, however, four out of the first nine subjects did not have activation in that area. If you bring each of these subjects back six months later to perform the same task, their particular pattern of response is consistent, but the variation between people remains high. So how can you apply group patterns to an individual?

There are also variations in how our brains are connected. The white matter in the brain, long neglected by science, is a vast network of fibers connecting neural structures. The way in which the brain processes information is dependent on how these fibers are connected. With diffusion tensor imaging (DTI) we can now look for individual variation in connections and it is proving to be tremendous.[21] Using DTI we have found that the way one person's corpus callosum is hooked up may be very different from someone else. This was first made evident to us by work in our lab in which we were calling on two processes: one, a process we knew was present in the right hemisphere, which rotates an object in space, and another process, in the left hemisphere, which names an object. For example, if I show you an up-side-down boat, before you can name it, you first rotate it right-side-up in your right hemisphere. Next, you send the rotated image to the left hemisphere and the left hemisphere names the object, and then you say it ("Ah, boat"). What we noticed is that some people are fast at this and some are slow at it. We found that people who are fast at naming use one part of their corpus callosum to transfer the information, and the slow people used a totally different part to get the information to their speech center. So then we thought that perhaps anatomical differences could explain this. It turns out that people vary tremendously in the number of fibers present in different parts of their callosum and in what routes are used to process this problem.[22] Capturing all this variation against or for a particular case in a legal setting may prove impossible.

TOO LITTLE TOO SOON BUT WATCH OUT!

Currently the case against using scans in the courtroom is quite evident for several reasons: (a) As I described, all brains are different from one another. It becomes impossible to determine if a pattern of activity in an individual is normal or abnormal. (b) The mind, emotions, and the way we think constantly change. What is measured in the brain at the time of scanning doesn't reflect what was happening at the time of a crime. (c) Brains are sensitive to many factors that can alter scans: caffeine, tobacco, alcohol, drugs, fatigue, strategy, menstrual cycle, concomitant disease, nutritional state, and so forth. (d) Performance is not consistent. People do better or worse at any task from day to day. (e) Images of the brain are prejudicial. A picture creates a bias of clinical certainty, when no such certainty is actually present. There are many firm reasons why in 2010, while I write this, although the science is enormously promising, it currently is not good enough, and it would more likely be misused instead of used properly. What we must remember, however, is things are changing fast in neuroscience and new technology is constantly allowing us to learn more about our brains and behavior. We have to be prepared for what may be coming in the future.

What may be coming has its foundation in the central principle in American criminal and common law, which is Sir Edward Coke's maxim of mens rea: The act does not make a person guilty unless the mind is also guilty. You need a guilty mind. Mens rea has four major parts that have to be demonstrated: (a) acting with the conscious purpose to engage in specific conduct or to cause a specific result (purposefulness); (b) awareness that one's conduct is of a particular nature, for instance, good or bad, legal or illegal (knowledge); (c) conscious disregard for a substantial and unjustifiable risk (recklessness); (d) the creation of a substantial or known risk of which one ought to have been aware (negligence). Each of those parts has brain mechanisms that are well studied and are still being studied. Purposefulness involves the brain's intentional systems; knowledge and awareness involves its emotional systems; recklessness involves the reward systems; and negligence involves joy-seeking systems. Much is already known about these areas, which will be causing problems for the principle of mens rea.

DONE BEFORE YOU KNOW IT?

As I mentioned in an earlier chapter, both the work of Benjamin Libet and Chun Siong Soon reveal that much of the work of the brain is done on the unconscious level and that a decision can be predicted several seconds before a subject consciously decides. The study of intention has become increasingly more interesting and has had some surprising and counter intuitive findings. If you take a normal person and stimulate the right parietal area at a low rate, the subject has the sensation that he has a conscious intention (I will lift my hand). If you stimulate at a higher rate a slightly different area in the parietal lobe, the subject has the awareness of action despite the fact that there has been no muscle action, that is, the subject hasn't done anything, but he believes that he has ("I have lifted my hand." Ah . . . no you didn't).[23] If, however, you stimulate the frontal area, he produces a multi-joint movement, but he has no awareness of it! From these studies it seems that it is the unconscious and not the conscious brain that is calling the shots. But hold on! While studies like these have spotlighted the "what" and "when" of intention, Marcel Brass and Patrick Haggard have begun to study one aspect of intention that has been oddly neglected: the "whether"[24] to implement the intention, the brakes that can be consciously thrown on that unconscious bubbling up. Their data suggest that a specific area in the dorsal fronto-median cortex (dFMC) is related to a kind of self-control[25] and have identified connectivity between it and motor preparation areas, which suggests that this self-control is achieved by modulating brain areas involved in motor preparation.[26] Individual differences among people in dFMC activation correlated with the frequency of inhibiting actions, and suggest a trait-like predisposition for self-control. They suggest that this is an example of top-down processing, where one mental state influences the next, and argues against hard determinism.

What we think of as willed activity has various components that can be separated into different brain areas, each of which can be identified. It is now understandable that when a brain scan is brought into the courtroom, if there is a lesion anywhere along the pathway from intention to action, a claim could be made that the person is either functioning normally or not. The scan, however, provides evidence of neither.

MIND READING

Mental states are important for determining guilt or innocence. In the future, increasing knowledge about mental states is going to lead to tighter claims about them and will have an enormous influence on how we think about ourselves and how the law will deal with this increased knowledge. *Mind reading*, which is actually detecting mental states, is a hot potato idea. The good old garden variety mind reader, lie detecting, has traditionally employed the notoriously unreliable polygraph test, which is only allowed in New Mexican courts and nowhere else in the United States. There are some new kids on the block that use EEG technology that have been admitted as evidence: Brain Fingerprinting in an Iowa courtroom in 2001[*] and in 2007 a court in India gave permission for two suspects in a murder to undergo a Brain Electrical Oscillations Signature test after a positive Polygraph Test was done. The positive results of this test were admitted as evidence in a trial in Pune, India, in 2008[†] that resulted in a murder conviction. Another new method using fMRI scanning (developed by the companies No Lie MRI and Cephos) has yet to appear in court. Many critics claim that there are not enough data to call any of these methods reliable. No test is infallible, and a certain percentage of falsely positive tests and falsely negative tests are consistently present in any given number of samples and determine how accurate a test result is. One can trust a test more if it is known that out of a thousand tests only 2 will be falsely positive rather than if 200 are falsely positive. For the above tests, the base rate of falsely positive and negative tests is not known. University of Virginia law professor Frederick Schauer[27] disagrees that these tests are not ready for prime time, arguing that science assumes the standards for law and science are the same, which they are not. He points out that the law's goals and

---

[*]Harrington v. State, 659NW 2nd 509 (Supreme Court Iowa 2003).
[†] http://lawandbiosciences.files.wordpress.com/2008/12/beosruling2.pdf

science's goals are quite different: While the prosecution has the heavier burden to prove guilt beyond a reasonable doubt, much like science requires for reliable data, the defense has to offer only reasonable doubt, and that is what some of these tests may provide, even if they don't have good reliability. He also points out the reliability and credibility of a self-interested witness are not good either. Currently the judge and jury determine when witnesses are telling the truth or lying, but the ordinary person's ability to spot liars is no better than random chance.[28]

Another mental state that can come under scrutiny in the courtroom is pain. Good methods of pain detection could separate malingerers from those who are really suffering in tort, disability and workmen's compensation cases. Detecting the conscious mental state in the absence of outward signs is also an active area of recent research and will determine decisions about withdrawing life support. While no test currently is reliable in detecting these mental states, they are on the horizon.

Ethical problems and legal problems, of course, are rampant. Is taking such a test equivalent to being a witness against oneself? Can the police get a warrant to read your mind? Is that invasion of privacy? What will the court's position be on those who refuse? When reliable, should tests be required in cases involving pain evaluation, disputing parties, on all witnesses, and so forth?

BIAS IN THE COURTROOM: JUDGES, JURORS AND ATTORNEYS

Supreme Court Justice Anthony Kennedy once said, "The law makes a promise: neutrality. If the promise gets broken, the law as we know it ceases to exist." Is neutrality even possible?

When a soldier in a war movie describes the enemy as all looking alike, he makes the hackles rise of the politically correct, and also is reflecting two unconscious brain processes, present in everyone including the politically correct, that can bias courtroom proceedings. One, the own-race bias (ORB) phenomenon, involves memory for human faces and has been widely reported in the psychological literature for more than seventy years. People are better able to correctly recognize face exemplars from their own-race compared with those from another racial group, and this phenomenon is not related to the level of prejudice. In a nation of great ethnic multiplicity, our weaker recognition of other-race faces is significant, and in fact, studies during the last twenty years, have revealed an increase in false positives: misidentifying someone as having been previously seen when they had not been.[29] This is of prime importance in the courtroom when it leads to the erroneous identification of someone who is not the perpetrator. In 1996 the U.S. Department of Justice reported that 85 percent of convictions that had been later overturned because of subsequent DNA analysis were due to erroneous eyewitness identifications.[30] One of the factors affecting the accuracy of other-race identification is "study time;" false alarms increase with shorter study time of the face, and eyewitnesses often only catch a quick glimpse. Accuracy also suffers with increasing time between viewing the crime and viewing a suspect.

This phenomenon has been utilized by expert witnesses and defense attorneys to dispute the efficacy of cross-race identification in the courtroom. While many theories about ORB abound, the simplest is that it is related to the frequency with which the perceiver encounters own-race faces relative to cross-race faces. A white kid growing up in Tokyo is going to be better at identifying Asian faces than a white kid growing up in Kansas. Knowing that development of perceptual expertise has been linked to the right brain, as has facial identification of others, one of my colleagues, David Turk, of the University of Aberdeen, wondered if the right brain too was where own-race processing was superior. He has now identified that while the right hemisphere is better at identifying faces in general, it is also better at own-race identification than other-race identification, whereas there is no difference in the lesser abilities of the left hemisphere. The own-race bias processes are localized to the right brain.[31] Now that there is a neurobiological basis to this bias, it can lead to the development of powerful tools for questioning witnesses and prospective jurors, and is another example of how neuroscience is going to be impacting the nature of evidence and ultimately the law.

The other unconscious brain process that may bias proceedings, dehumanizing out-groups, has been studied by Lasana Harris and Susan Fiske.[32] They found that, when American subjects view certain social groups, different emotions are elicited depending on what group it is. The emotions of envy (when viewing the rich), pride (seeing American Olympic athletes), and pity (while viewing photos of elderly people) are all associated with activity in the area of the brain (the medial prefrontal cortex, or mPFC) that is activated in social encounters. However, the emotion of disgust (looking at photos of drug addicts) is not. The activation patterns in the mPFC while viewing photos of social groups that elicited disgust were no different when the subjects viewed objects, such as a rock. This suggests that members of groups that elicit disgust, which are extreme out-groups, are dehumanized. This is what occurs during war: the enemy group elicits disgust and is dehumanized and pejoratively labeled. Jurors, judges, and attorneys all have unconscious neural responses to certain people that can powerfully influence their behavior and potentially change how a person will be evaluated. The legal system has heeded the findings of such studies and is not blind to the influences of unconscious bias. Attorneys are constantly looking for bias while selecting jurors, and those who are selected are warned to guard against it, in an appeal for top-down processing by the judge.

GUILTY AS CHARGED:
TO PUNISH OR NOT TO PUNISH?

> If you had come to me in friendship, then this scum that
> ruined your daughter would be suffering this very day.
> —*The Godfather*

In the court systems, however, complicated as they may be, proceedings that arrive at a verdict are the easy part. Most of the defendants that get to trial or plead guilty are the agents of the crime. After a defendant has been found guilty, next comes the sentencing. That is the hard part. What do you do with the guilty, who have intentionally committed known, morally wrong actions that harm others? In the United States, if you are an offender in a criminal law case, you face "punishment," whereas if you fall under the jurisdiction of civil law, the goal is for the offender to compensate the injured party. The judge looks at all the mitigating and contributing factors (age, previous criminal record, severity of the crime, negligence versus intention, unforeseeable versus foreseeable harm, and so forth), sentencing guidelines, and then makes a decision.

That decision is supposed to mete out justice, and therein lies the rub. Justice is a concept of moral rightness, but there has never been an agreement as to what moral rightness is based on: ethics (should the punishment fit the crime, retribution, or be for the greater good of the population, utilitarian?), reason (will punishment or treatment lead to a better outcome?), law (a system of rules that one agrees to live by in order to maintain a place in society), natural law (actions results in consequences), fairness (based on rights? based on equality or merit? based on the individual or society?), religion (based on which one?), or equity (allowing the court to use some discretion over sentencing)? Nonetheless, the judge tries to come up with a just disposition. Should the offender be punished? If so, should the goal of punishment be mindful of individual rights based on retribution, mindful of the good of society with reform and deterrence in mind, or mindful of the victim with compensation? This decision is affected by the judge's own beliefs of justice, which come in three flavors: retributive justice, utilitarian justice, and the up-and-coming restorative justice.

Retributive justice is backward-looking. One is punished in proportion to the crime that is committed, extending just deserts to the individual, and punishment is the goal. The crucial variable is the degree of moral outrage the crime engenders, not the benefits to society resulting from the punishment. Therefore, one does not get a life sentence for stealing a CD player, nor does one get a month's probation for murder. One does not get punishment if one is judged insane. The punishment is focused solely on what the individual deserves for his crime, not more or less. It appeals to the intuitive sense of fairness where every individual is equal and is punished equally. You cannot be punished for crimes you have not committed. You don't get a higher fine because you are rich or a lower fine because you are poor. No matter who you are, you should receive the same punishment. You do not get a harsher sentence because you are or are not famous, because you are black or white or brown. It is not part of a calculation for the general welfare of society as a whole. Retributive justice does not punish as a deterrent to others, to reform the offender, or to compensate the victim. These may result as by-products, but they are not the goal. It punishes to harm the offender, just as the victim was harmed.

Utilitarian justice (consequentialism) is forward looking and concerned about the greater future good of society resulting from punishing the individual offender. This is accomplished by assigning one of three types of punishment. The first will specifically *deter* the offender (or others that will learn by example) in the future, perhaps by fines, prison time or community service. The second type will incapacitate him. Incapacitation can be achieved geographically, by long prison sentences or banishment, which includes disbarment for lawyers and other such licensing losses, or by physical means, such as chemical castration for rapists and capital punishment. The third type of utilitarian justice is rehabilitation through treatment or education. The method chosen is decided by the probability of recidivism, degree of impulsiveness, criminal record, ethics (can treatment be forced upon someone who is unwilling to undergo it?) and so forth, or by prescribed sentencing standards. This is another area where neuroscience will have something to contribute. Prediction of future criminal behavior is pertinent to utilitarian sentencing decisions, whether treatment, probation, involuntary commitment or detention is chosen. Neural markers could be used to help identify an individual as a psychopath, sexual predator, impulsive, and so forth, in conjunction with other evidence to make predictions of future behavior. Obviously the reliability of such predictions is important, remembering that and utilitarian justice punishes for uncommitted future crimes, and can result in either decreasing or increasing harmful errors.

Utilitarian justice also may punish one person to deter others, the severity need not relate to the actual offence: A thief of a CD player could receive a harsh sentence to deter others from thieving. Thus, it makes sense to punish a famous person or the perpetrator of a highly publicized crime more harshly, because the publicity may deter many future crimes and benefit society. Arguments have been made from the utilitarian standpoint that it makes sense to have harsh sentences for the more common milder offences to increase the deterrence effect. Prison sentences for first-time speeding and drunk driving offences may save more innocent lives than punishing convicted murders. The extreme case can even be made that the punished need not even be guilty, just thought guilty by the public. An innocent person could be arrested as a scapegoat and their imprisonment could stave off a vigilante effort or riot for the greater good. This is why utilitarian justice can rub people wrong, it can violate an individual's rights, it may not seem "fair."

Restorative justice looks at crimes as having been committed against a person rather than against the state. While this focus on persons was common in the ancient cultures of Babylon, Sumer, and Rome, this all changed with the Norman invasion of Great Britain in (wasn't this date drilled into your head in high school?) 1066. William the Conqueror, in an effort to centralize power, saw crime as an injury to the state, where the victim had no part in the justice system. This viewpoint is also seen to insure neutrality in criminal proceedings, avoiding vengeful and unfair retaliation, and it remained the prominent or dominant view in American law until late in the twentieth century. In 1974, a Mennonite probation officer and a volunteer service director in Kitchener, Ontario, Canada, began a discussion group looking for ways to improve the criminal justice system, and a recent version of restorative justice was born, now with varied versions. It focuses on the needs of both the victim and the offender. It attempts to repair the harm done to the victim and to make the victim whole, and it attempts to make the offender law-abiding in society.

Restorative justice holds the offender directly accountable to the victim and the affected community, requires the offender to make things whole again to the extent that it is possible, allows the victim a say in the corrections process, and encourages the community to hold offenders accountable, to support victims, and to provide opportunities for offenders to reintegrate themselves into the community.[33] Victims, offenders, and the community play an active role. Victims of crimes often are enveloped in fear, adversely affecting the rest of their lives, as Mary Vincent was at the beginning of this chapter. This can be true for whole communities also. For crimes of lesser magnitude, often a face to face sincere apology and reparation are enough to relieve the victim of their fear and anger. Restorative justice may not be possible for more serious crimes.

WE ARE JUDGE AND JURY FROM BIRTH

Although judges, juries, and attorneys most likely will attribute their stances to various factors, not the least of which are long years of education, philosophical discussion and the like, as usual, most of the goings on in the courtroom are intuitions that we came with from the baby factory, including a sense of fairness, reciprocity, and punishment. Renee Baillargeon and colleagues have been hard at work with a group of toddlers and have shown that a sense of fairness is present not only in two-and-a-half-year-olds, but also sixteen-month-olds. The older group when asked to distribute treats to animated puppets will do so evenly,[34] and the sixteen-month-old infants prefer cartoon characters that divide prizes equally.[35] We also come wired for reciprocity, but only within our social group. Toddlers expect members of a group to play and share toys,[36] and are surprised when it doesn't happen. They are not surprised when it doesn't happen between members of different groups but surprised when it does.

The toddlers in Michael Tomasello's lab not only recognize moral transgressors, but react negatively to them. One-and-a-half- to two-year-olds help, comfort, and share with a victim of a moral transgression, even in the absence of overt emotional cues. With the perps, it is another story. Moral transgressors incite the infants to vocally protest and they are less inclined to help, comfort, or share with them.[37] Young children also understand intentionality and judge intentional violations of rules as "naughty" but not accidental ones.[38] While it is well known that adults will willingly suffer to punish others, a yet-to-be-published study of Paul Bloom's lab has shown that this is even true for four-year-olds.[39] We feel these urges all the time; we try to have big theories about them, but we are just born that way.

NOT PUTTING YOUR MONEY
WHERE YOUR MOUTH IS

What people say they believe about punishment and what their actual behavior is are two different animals, and they aren't really able to offer logical explanations why. We have run into this before, haven't we? The interpreter is back at work trying to explain an intuitive judgment. Psychology graduate student Kevin Carlsmith and his advisor, John Darley, were curious. When people were asked to label themselves as retributivists or deterrists, their answers varied widely, and they divided themselves into either one of the two groups or into a third group and labeled themselves mixed. These individual differences, however, only mildly affected their punitive behavior, which was retributivist for the most part. They found when people are given a task to assign hypothetical punishment for an offense, 97 percent seek out information relevant to a retributive perspective and not to the utilitarian perspective (incapacitation or deterrence).[40] They are highly sensitive to the severity of the offense and ignore the likelihood that the person would offend again. They punish for the harm done, not for the harm that might be done in the future (deterrence). When asked to punish only from the utilitarian perspective and to ignore retributive factors, which had carefully been explained to them, they still did not. People still used the severity of the crime to guide their judgments.[41] When they are forced to take the utilitarian perspective, they feel less confident in their decisions. When asked to allocate resources for catching offenders or preventing crime, they highly supported the utilitarian approach of preventing crime. So although people endorse the utilitarian theory of reducing crime, they don't want to do it through unjust punishment. They want to give a person what they deserve, but only after they deserve it. They want to be fair. "[P]eople want punishment to incapacitate and to deter, but their sense of justice requires sentences proportional to the moral severity of the crime."[42] (Even the Catholic Church makes the distinction with the more light weight venal sins being punished by time in purgatory, while mortal sins send you straight to hell.) This appeal to fairness goes along with the finding that people give lighter hypothetical punishments after reading about determinism. If offenders aren't responsible for their actions, then they don't deserve harsh punishment.

The reasons people give for their punishments, however, do not match what they do. They endorse utilitarian policies in the abstract, but invoke retributivist ones in practice.[43] Carlsmith and Darly point out that this lack of insight leads to fickle legislation. For instance, 72 percent of the voters of California enacted the three-strikes law that put a person convicted of a third felony in prison for life, a utilitarian approach. A few years later when people realized that this could mean an "unfair" life sentence for stealing a piece of pizza, support dropped to less than 50 percent, sensing that the law was unfair from a retributivist perspective. Because of this highly intuitive "just deserts" impulse, these authors suggest that when considering the idea of restorative justice, which is appealing, they doubt that citizens will be willing to allow a purely restorative, punishment-absent treatment for serious crimes. In a scenario where people could choose to assign cases to various court systems, restorative only, retributivist only, or a combination, 80 percent were willing to send minor crimes to restorative courts, but only 10 percent elected restorative courts for serious crimes, while 65 percent opted for mixed and 25 percent to retributivist courts. It appears we share the same moral response to punishment. As we saw in the last chapter with other moral systems, the only thing that is different is not our behavior but our theories about why we respond the way we do.

If a judge holds the belief that people are personally responsible for their behavior, then either retributive punishment or restorative justice makes sense; if the judge believes that deterrence is effective, or that punishment can change bad behavior into good, or that some people are irredeemable, then utilitarian punishment makes sense; if the judge has a determinist stance, then there is a decision to be made. Either his focus of concern will be for: (1) the offender's individual rights and because the offender had no control over his determined behavior, he or she should not be punished but perhaps should be treated (but not against their will?) if possible, or (2) for the victim's rights of restitution and any deterministic retributive feelings the victim might have, or for (3) the greater good of society (it may not be the offenders' fault, but get 'em off the streets).

NOTHING NEW UNDER THE SUN

The sun, as it glides over Athens, is no doubt yawning and rolling his (being over Athens we're talking Apollo here) eyes . . . "Haven't they got this thing settled yet? I have been hearing this same old argument for century after century." Aristotle argued that justice based on fair treatment of the individual leads to a fair society, whereas Plato, looking at the big picture, thought fairness to society was of primary importance and individual cases were judged in order to achieve that end. It is back to a version of the dichotomy found between Western and East Asian thought: where should we place our attention, on the individual or the community?

These two ways of thinking also take us back to the trolley problem: the emotional situation and the more abstract situation. Facing the individual offender in a courtroom and deciding whether to punish is an emotional proposition, and elicits an intuitive emotional reaction: "Throw the book at 'em!" or, "Poor guy, he didn't mean to do it, let him off easy!" In a recent fMRI[44] study done while subjects were judging responsibility and assigning punishment in hypothetical cases, brain regions associated with emotion activated during the punishment judgment, the more activity the greater the punishment (as with retribution, the greater the moral outrage, the greater the punishment). The region of the right dorsolateral prefrontal cortex that is recruited when judgments about punishments are made in the ultimatum economic game corresponds to that which was recruited while making third party legal decisions. These researchers suggest that "our modern legal system may have evolved by building on pre-existing cognitive mechanisms that support fairness-related behaviors in dyadic interactions." If an evolutionary link to relations between individuals in socially significant situations (for example, mates) is true, it makes sense that faced with an individual we resort to fairness judgments, rather than consequentialist. Faced, however, with the abstract questions of public policy, then we leave the emotional reaction behind and can resort to the more abstract consequentialist thinking.

Philosopher Janet Radcliffe Richards stated,

. . . many people accept that the arguments about free will and ultimate responsibility really do show that no one can ultimately deserve punishment . . . If so, punishment cannot be justified on the retributivist grounds that it is ultimately deserved, but only on the consequentialist grounds that it is necessary for deterring antisocial behavior.

. . . if we understand that there are good *evolutionary* reasons for our wanting people to suffer when they have done direct or indirect harm to us, then we can account for our strong feelings about the appropriateness of retribution without presuming they are a guide to moral truth. . . . We may be able to recognize our retributivist feelings as a deep and important aspect of our character— and take them seriously to that extent—without endorsing them as a guide to truth, and start rethinking our attitudes toward punishment on that basis.[45]

She goes on to say, however, that she has no idea how to go about it.

DELICATE BALANCES:
CAN A SOCIETY BE CIVILIZED
AND LIVE WITH PUNISHMENT?

Will the system work without punishment? This is a stance that is advocated by the hard core determinists such as Boalt law professor Sanford Kadish, who has written, "To blame a person is to express moral criticism, and if the person's action does not deserve criticism, blaming him is a kind of falsehood, and is, to the extent the person is injured by being blamed, unjust to him." Actually, one can interpret this stance as coming from a retributivist viewpoint. If one has no control over one's determinist brain then one doesn't deserve punishment, a retributivist argument. The same can be said for the conclusion reached in the court decision of Holloway vs. U.S. in 1945: "To punish a man who lacks the power to reason is as undignified and unworthy as punishing an inanimate object or animal. A man who cannot reason cannot be subject to blame." It could just as easily have said it is not fair to punish someone who does not deserve it. Is forgiveness a viable concept? Is it possible to run a society where forgiveness trumps accountability and punishment? Would such a system work?

As I mentioned in the last chapter, unlike any other species, we humans have evolved to cooperate on a massive scale with unrelated others. This has been difficult to explain from the evolutionary standpoint because cooperating individuals incur costs to themselves that benefit non-kin, which doesn't make sense at the individual level. How can that be a strategy for success? The reason is that it does make sense on the group level. We have seen that in the ultimatum game people will punish noncooperators at personal cost even in one shot games. It turns out that both theoretical models and experimental evidence show that in the absence of punishment, cooperation both in large and small groups cannot sustain itself in the presence of free-riders, and collapses.[46] In order for cooperation to survive, free-riders must be punished. If you take accountability out of the network, the whole thing falls apart. Can you have accountability without punishment? Clearly our genome thinks it is important. Can we or should we rise above it or not? Punishing free riders in economic games or those that don't follow the accepted rules of a social group, brings us back again to Tomasello's theory of self domestication of humans: Punishment by incapacitation (whether it was by killing or banishment) resulted in temperaments being selected for that made us more cooperative. If we don't incapacitate the offenders, will the noncooperators take over and society fall apart?

All these issues are being generated by a more physicalist understanding about who we are and that understanding is, in turn, going to influence how we think about the issues. There are problems on both sides.

SOCIAL INTERACTIONS MAKES US FREE TO CHOOSE

My contention is that ultimately responsibility is a contract between two people rather than a property of a brain, and determinism has no meaning in this context. Human nature remains constant, but out in the social world behavior can change. Brakes can be put on unconscious intentions. I won't throw my fork at you because you took a bite of my biscuit. The behavior of one person can affect another person's behavior. I see the highway patrolman coming down the onramp and I check my speedometer and slow down. As I said in the last chapter, the point is that we now understand that we have to look at the whole picture, a brain in the midst of and interacting with other brains, not just one brain in isolation.

No matter what their condition, however, most humans can follow rules. Criminals can follow the rules. They don't commit their crimes in front of policemen. They are able to inhibit their intentions when the cop walks by. They have made a choice based on their experience. This is what makes us responsible agents, or not.

Chapter Seven

AN AFTERWORD

I REMEMBER WATCHING A HAUNTING BBC DOCUMENTARY A few years back that told a simple story. An experienced BBC reporter was visiting India and decided to look up an Indian friend. The film rolled on with the cameraman and reporter slogging through streets of muck and excrement in a hillside shantytown to his friend's 8 × 10 foot home. There he was, smiling and beaming at seeing his pal from the U.K. It turned out his home, where he lived with his wife and two children, was also his place of work and his shop. He sold children's tennis shoes, the kind that blink. Somehow they made it all work in this small place, and, while the cameraman was itching to leave as he could not bear the smells, the dignified Indian handed his English friend a pair of shoes to take back to his kids. Here they were in what a Westerner would only call abject poverty and misery, and yet the human exchange transcended everything—that moment that so defines who we are. It is that magnificence of being "human" that we all cherish and love and that we don't want science to take away. We want to feel our own worth and the worth of others.

I have tried to argue that a more complete scientific understanding of the nature of life, of brain/mind is not eroding this value we all hold dear. We are people, not brains. We are that abstraction that occurs when a mind, which emerges from a brain, interacts with the brain. It is in that abstraction that we exist and in the face of science seeming to chip away at it, we are desperately seeking a vocabulary to describe what it is we truly are. We all are endlessly curious about how it all works. The large deterministic view that surrounds all of science seems to be urging a more bleak view, the view that no matter how we dress it up, in the end we are machines of some kind, automatically and mindlessly serving as the vehicles for the physically determined forces of the universe, forces larger than us. Each of us is not precious. We are all pawns.

The common way out of this dilemma is to ignore it and to say something about how great life is at a phenomenological level, how beautiful Yosemite is, how wonderful sex is and grandchildren, too, and to groove on it all. We groove because we are built to enjoy these things. That is the way we work, and that is the end of the matter. Go have a dry martini, put your feet up, and read a good book.

I have tried to offer a different perspective on this dilemma. In the end, my argument is that all of life's experiences, personal and social, impact our emergent mental system. These experiences are powerful forces modulating the mind. They not only constrain our brains but also reveal that it is the interaction of the two layers of brain and mind that provides our conscious reality, our moment in real time. Demystifying the brain is the task of modern neuroscience. To complete that job, however, will require neuroscience to think about how the rules and algorithms that govern all of the separate and distributed modules work together to yield the human condition.

Understanding that the brain works automatically and follows the laws of the natural world is both heartening and revealing. Heartening because we can be confident the decision-making device, the brain, has a reliable structure in place to execute decisions for actions. It is also revealing, because it makes clear that the whole arcane issue about free will is a miscast concept, based on social and psychological beliefs held at particular times in human history that have not been borne out and/or are at odds with modern scientific knowledge about the nature of our universe. As John Doyle has put it to me:

> Somehow we got used to the idea that when a system appears to exhibit coherent, integrated function and behavior, there must be some "essential" and, importantly, central or centralized controlling element that is responsible. We are deeply essentialist, and our left brain will find it. And as you point out, we'll make up something if we can't find it. We call it a homunculus, mind, soul, gene, etc. . . . But it is rarely there in the usual reductionist sense. . . . That doesn't mean there isn't in fact some "essence" that is responsible, it's just distributed. It's in the protocols, the rules, the algorithms, the software. It's how cells, ant hills, Internets, armies, brains, really work. It's difficult for us because it doesn't reside in some box somewhere, indeed it would be a design flaw if it did because that box would be a single point of failure. It's, in fact, important that it not be in the modules but in the rules that they must obey.

As I wind down this effort, I find my own perspective adjusting. That is the nature of a life in science. The facts don't change. What changes, especially in highly interpretive sciences such as neuroscience and psychology, are the ideas on how to understand the ever-accumulating facts of Mother Nature. Every morning, the gnawing question every scientist asks again and again is: Does that explanation I have for such and such really capture what is going on? No one knows more about the weaknesses in one's idea than the person proposing it, and as a consequence, one is always on the lookout. This is not a particularly easy state to be in, and I once asked Leon Festinger, one of the smartest men in the world, whether or not he ever felt inept. He replied, "Of course! That is what keeps you ept."

While reviewing material for this book, I realized that a unique language, which has yet to be developed, is needed to capture the thing that happens when mental processes constrain the brain and vice versa. The action is at the interface of those layers. In one kind of vocabulary it is where downward causation meets upward causation. In another vocabulary it is not there at all but in the space between brains that are interacting with each other. It is what happens at the interface of our layered hierarchical existence that holds the answer to our quest for understanding mind/brain relationships. How are we to describe that? That emergent level has its own time course and is current with the actions taking place. It is that abstraction that makes us current in time, real and responsible. The whole business about the brain doing it before we are conscious of it becomes moot and inconsequential from the vantage point of a different level of operation. Understanding how to develop a vocabulary for those layered interactions, for me, constitutes the scientific problem of this century.

ACKNOWLEDGMENTS

My debt to colleagues, family, and institutions continues to grow with every book I write. In this case, the University of Edinburgh and the Gifford Lecture series served as the catalyst. I was humbled by the invitation and exhilarated by the challenge to give a series of lectures over two weeks in the fall of 2009. My goal was to articulate what I thought neuroscience has taught us about some of the great philosophical issues of life and in particular what it has taught us about being held responsible for our actions. Lots of people want to know about this, including, to my surprise, Charlotte, my wife; my kids, Marin, Anne, Francesca, and Zachary; my son-in-law, Chris; and my sister, Rebecca. They all showed up in Edinburgh, rented an apartment, and held my feet to the fire. It was a magnificent time, or so they tell me. Needless to say I was anguishing over the lectures.

Of course, in the large scheme of things, lecturing is the easy part. While it does prompt one to pull their thoughts together, writing out one's thoughts is another matter. Once again, I have been aided by many people. My sister, Rebecca, has become indispensible. Her editorial hand and wit has broadened my own propensity for the conversational approach to communicating. I can't thank her enough. My deep gratitude also goes to Jane Nevins, my friend and colleague from the Dana Foundation. Her razor eye and firm editing is second to none. She leaves your style alone and only comes after you when you are dead wrong. That happens too frequently for my taste, but I learn every time.

Thanking all my professional colleagues would be impossible. Over the years I have been inspired by many, starting with my mentor, Roger Sperry, perhaps the greatest brain scientist who ever lived. One can also see from the perspective in this book I have been heavily influenced by my many graduate and postdoctoral students. They are as much a part of this research and perspective as I am. There have also been some giants in the field such as Leon Festinger, George Miller, and David Premack, who have tried their best to make me better than I am. Former Gifford Lecturer Donald Mackay did the same. So too for Michael Posner, Steven Hillyard, Leo Chalupa, Floyd Bloom, Emilio Bizzi, Marc Raichle, Scott Grafton, Endel Tulving, Steve Pinker, and many, many more. It has been a rich life. Special thanks also to Walter Sinnott-Armstrong and Michael Posner for their critique of this manuscript, and to John Doyle at Caltech for his reading and endless insights as to where the field of mind/brain research must go in the future. I started my career at Caltech, and it is gratifying to be back knocking at their door to learn more.

NOTES

CHAPTER 1: THE WAY WE ARE

1. Hippocrates (400 B.C.). Hippocratic writings (Francis Adams, Trans.). In M. J. Adler (Ed.), *The great books of the western world* (1952 ed., Vol. 10, p. 159). Chicago: Encyclopædia Britannica, Inc.
2. Doyle, A. C. (1892). Silver blaze. In *The complete Sherlock Holmes* (1930 ed., Vol. 1, p. 335). Garden City, NY: Doubleday & Company, Inc.
3. Lashley, K. S. (1929). *Brain mechanisms and intelligence: A quantitative study of injuries to the brain.* Chicago: University of Chicago Press.
4. Watson, J. B. (1930). *Behaviorism* (Rev. ed., p. 82). Chicago: University of Chicago Press.
5. Weiss, P. A. (1934). In vitro experiments on the factors determining the course of the outgrowing nerve fiber. *Journal of Experimental Zoology, 68*(3), 393–448.
6. Sperry, R. W. (1963). Chemoaffinity in the orderly growth of nerve fiber patterns and connections. *Proceedings of the National Academy of Sciences of the United States of America, 50*(4), 703–710.
7. Hebb, D. O. (1949). *The organization of behavior: A neuropsychological theory* (p. 62). New York: Wiley.
8. Hebb, D. O. (1947). The effects of early experience on problem solving at maturity. *American Psychologist, 2*, 306–307.
9. Ford, F. R., & Woodall, B. (1938). Phenomena due to misdirection of regenerating fibers of cranial, spinal and autonomic nerves. *Archives of Surgery, 36*(3), 480–496.
10. Sperry, R. (1939). The functional results of muscle transposition in the hind limb of the rat. *The Journal of Comparative Neurology, 73*(3), 379-404.
11. Sperry, R. (1943). Functional results of crossing sensory nerves in the rat. *The Journal of Comparative Neurology, 78*(1), 59–90.

12. Sperry, R. W. (1963). Chemoaffinity in the orderly growth of nerve fiber patterns (p. 703).

13. Pomerat, C. M. (1963). Activities associated with neuronal regeneration. *The Anatomical Record, 145*(2), 371.

14. Krubitzer, L. (2009). In search of a unifying theory of complex brain evolution. *Annals of the New York Academy of Science, 1156,* 44–67.

15. Marler, P., & Tamura, M. (1964). Culturally transmitted patterns of vocal behavior in sparrows. *Science, 146*(3650), 1483–1486.

16. Jerne, N. (1967). Antibodies and learning: selection versus instruction. *The neurosciences: A study program* (pp. 200–205) New York: Rockefeller University Press.

17. Boag, P. T., & Grant, P. R. (1981). Intense natural selection in a population of Darwin's Finches (Geospizinae) in the Galápagos. *Science, 214*(4516), 82–85.

18. Sin, W. C., Haas, K, Ruthazer, E. S., & Cline, H. T. (2002). Dendrite growth increased by visual activity requires NMDA receptor and Rho GTPases. *Nature, 419*(6906), 475–480.

19. Rioult-Pedotti, M. S., Donoghue, J. P., & Dunaevsky, A. (2007). Plasticity of the synaptic modification range. *Journal of Neurophysiology, 98*(6), 3688–3695.

20. Xu, T., Yu, X., Perlik, A. J., Tobin, W. F., Zweig, J. A., Tennant, K., . . . Zuo, Y. (2009). Rapid formation and selective stabilization of synapses for enduring motor memories. *Nature, 462*(7275), 915–919.

21. Baillargeon, R. E. (1987). Object permanence in 3½ and 4½ month old infants. *Developmental Psychology, 23*(5), 655–664.

22. See: Spelke, E. S. (1991). Physical knowledge in infancy: Reflections on Piaget's theory. In S. Carey & R. Gelman (Eds.), *The epigenesis of mind: Essays on biology and cognition* (pp. 133–169). Hillsdale, NJ: Lawrence Erlbaum Associates; and Spelke, E. S. (1994). Initial knowledge: Six suggestions. *Cognition, 50,* 443–447.

23. Purves, D., Williams, S. M., Nundy, S., & Lotto, R. B. (2004). Perceiving the intensity of light. *Psychological Review, 111*(1), 142–158.

24. Purves, D. An empirical explanation: Simultaneous brightness contrast. Retrieved from: http://www.purveslab.net/research/explanation/brightness/brightness.html#f2.

25. Lovejoy, C. O., Latimer, B., Suwa, G., Asfaw, B., & White, T. D. (2009). Combining prehension and propulsion: The foot of Ardipithecus ramidus. *Science, 326*(5949), 72, 72e1–72e8.

26. Festinger, L. (1983). *The human legacy* (p. 4). New York: Columbia University Press.

27. Lovejoy, C. O. (2009). Reexamining human origins in light of Ardipithecus ramidus. *Science, 326*(5949), 74, 74e1–74e8.

28. Darwin, C. (1871). *The descent of man, and selection in relation to sex.* London: John Murray (Facsimile ed., 1981, Princeton, NJ: Princeton University Press).

29. Huxley, T. H. (1863). *Evidence as to man's place in nature.* London: Williams and Morgate (Reissued, 1959, Ann Arbor: University of Michigan Press).

30. Holloway, R. L., Jr. (1966). Cranial capacity and neuron number: A critique and proposal. *American Journal of Anthropology, 25*(3), 305–314.

31. Holloway, R. L. (2008). The human brain evolving: A personal retrospective. *Annual Review of Anthropology, 37,* 1–19.

32. See: Preuss, T. M., Qi, H., & Kaas, J. H. (1999). Distinctive compartmental organization of human primary visual cortex. *Proceedings of the National Academy of Sciences of the United States of America, 96*(20), 11601–11606; and Preuss, T. M., & Coleman, G. Q. (2002). Human-specific organization of primary visual cortex: Alternating compartments of dense cat-301 and calbindin immunoreactivity in layer 4A. *Cerebral Cortex, 12*(7), 671–691.

33. de Winter, W., & Oxnard, C. E. (2001). Evolutionary radiations and convergences in the structural organization of mammalian brains. *Nature, 409,* 710–714.

34. Oxnard, C. E. (2004). Brain evolution: Mammals, primates, chimpanzees, and humans. *International Journal of Primatology, 25*(5), 1127–1158.

35. Rakic, P. (2005). Vive la difference! *Neuron, 47*(3), 323–325.

36. Premack, D. (2007). Human and animal cognition: Continuity and discontinuity. *Proceedings of the National Academy of Sciences of the United States of America, 104*(35), 13861–13867.

37. Azevedo, F. A. C., Carvalho, L. R. B., Grinberg, L. T., Farfel, J. M., Ferretti, R. E. L., Leite, R. E. P., . . . Herculano-Houzel, S. (2009). Equal numbers of neuronal and nonneuronal cells make the human brain an isometrically scaled-up primate brain. *Journal of Comparative Neurology, 513*(5), 532–541.

38. Shariff G. A. (1953). Cell counts in the primate cerebral cortex. *Journal of Comparative Neurology, 98*(3), 381–400.

39. Deacon, T. W. (1990). Rethinking mammalian brain evolution. *American Zoology, 30*(3), 629–705.

40. Ringo, J. L. (1991). Neuronal interconnection as a function of brain size. *Brain, Behavior and Evolution, 38*(1) 1–6.

41. Petersen, S. E., Fox, P. T., Posner, M. I., Mintun, M., & Raichle, M. E. (1988). Positron emission tomographic studies of the cortical anatomy of single-word processing. *Nature, 331*(6157), 585–589.

42. Preuss, T. M. (2001). The discovery of cerebral diversity: An unwelcome

scientific revolution. In D. Falk & K. R. Gibson (Eds.), *Evolutionary anatomy of the primate cortex* (p. 154). Cambridge: Cambridge University Press.

43. Hutsler, J. J., Lee, D. G., & Porter, K. K. (2005). Comparative analysis of cortical layering and supragranular layer enlargement in rodent carnivore and primate species. *Brain Research, 1052,* 71–81.

44. See the following: Caviness, V. S., Jr., Takahashi, T., & Nowakowski, R. S. (1995). Numbers, time and neocortical neurogenesis: A general developmental and evolutionary model. *Trends in Neuroscience, 18*(9), 379–383; Fuster, J. M. (2003). Neurobiology of cortical networks. In *Cortex and mind* (pp. 17–53). New York: Oxford University Press; and Jones, E. G. (1981). Anatomy of cerebral cortex: Columnar input-output organization. In F. O. Schmitt, F. G. Worden, G. Adelman, & S. G. Dennis (Eds.), *The organization of the cerebral cortex* (pp. 199–235). Cambridge, MA: The MIT Press.

45. Hutsler, J. J., & Galuske, R. A. W. (2003). Hemispheric asymmetries in cerebral cortical networks. *Trends in Neuroscience, 26,* 429–435.

46. Elston, G. N., & Rosa, M. G. P. (2000). Pyramidal cells, patches and cortical columns: A comparative study of infragranular neurons in TEO, TE, and the superior temporal polysensory area of the macaque monkey. *The Journal of Neuroscience, 20*(24), RC117.

47. Elston, G. N. (2003). Cortex, cognition and the cell: New insights into the pyramidal neuron and prefrontal function. *Cerebral Cortex, 13*(11), 1124–1138.

48. Rilling, J. K., & Insel, T. R. (1999). Differential expansion of neural projection systems in primate brain evolution. *Neuroreport, 10*(7), 1453–1459.

49. See: Buxhoeveden, D., & Casanova, M. (2000). Comparative lateralisation patterns in the language area of human, chimpanzee, and rhesus monkey brains. *Laterality, 5*(4), 315–330; and Gilissen, E. (2001). Structural symmetries and asymmetries in human and chimpanzee brains. In D. Falk & K. R. Gibson (Eds.), *Evolutionary anatomy of the primate cerebral cortex* (pp. 187–215). Cambridge: Cambridge University Press.

50. Vermeire, B., & Hamilton, C. R. (1998). Inversion effect for faces in split-brain monkeys. *Neuropsychologia, 36*(10), 1003–1014.

51. Halpern, M. E., Güntürkün, O., Hopkins, W. D., & Rogers, L. J. (2005). Lateralization of the vertebrate brain: Taking the side of model systems. *Journal of Neuroscience, 25*(35), 10351–10357.

52. For a review, see: Hutsler, J. J., & Galuske, R. A. W. (2003). Hemispheric asymmetries in cerebral cortical networks. *Trends in Neuroscience, 26*(8), 429–435.

53. Black, P., & Myers, R. E. (1964). Visual function of the forebrain commissures in the chimpanzee. *Science, 146*(3645), 799–800.

54. Pasik, P., & Pasik, T. (1982). Visual functions in monkeys after total removal of visual cerebral cortex. In W. D. Neff (Ed.), *Contributions to sensory physiology* (Vol. 7, pp. 147–200). New York: Academic Press.

55. Rilling, J. K., Glasser, M. F., Preuss, T. M., Ma, X., Zhao, T., Hu, X., & Behrens, T. E. J. (2008). The evolution of the arcuate fasciculus revealed with comparative DTI. *Nature Neuroscience, 11*(4), 426–428.

56. Preuss, T. M. (2003). What is it like to be a human? In M. S. Gazzaniga (Ed.), *The Cognitive Neurosciences III* (pp. 14–15). Cambridge, MA: The MIT Press.

57. Elston, G. N. (2003). Cortex, cognition and the cell: New insights into the pyramidal neuron and prefrontal function. *Cerebral Cortex, 13*(11), 1124–1138.

58. Elston, G. N., Benavides-Piccione, R., Elston, A., Zietsch, B., Defelipe, J., Manger, P., . . . Kaas, J. H. (2006). Specializations of the granular prefrontal cortex of primates: Implications for cognitive processing. *The Anatomical Record, 288A*(1), 26–35.

59. Williamson, A., Spencer, D. D., & Shepherd, G. M. (1993). Comparison between the membrane and synaptic properties of human and rodent dentate granule cells. *Brain Research, 622*(1–2), 194–202.

60. Nimchinsky, E. A., Vogt, B. A., Morrison, J. H., & Hof, P. R. (1995). Spindle neurons of the human anterior cingulate cortex. *Journal of Comparative Neurology, 355*(1), 27–37.

61. Fajardo, C., Escobar, M. I, Buriticá, E., Arteaga, G., Umbarila, J., Casanova, M. F., & Pimienta, H. (2008) Von Economo neurons are present in the dorsolateral (dysgranular) prefrontal cortex of humans. *Neuroscience Letters, 435*(3), 215–218.

62. Nimchinsky, E. A., Gilissen, E., Allman, J. M., Perl, D. P., Erwin, J. M., & Hof, P. R. (1999). A neuronal morphologic type unique to humans and great apes. *Proceedings of the National Academy of Sciences of the United States of America, 96*(9), 5268–5273.

63. Allman, J. M., Watson, K. K., Tetreault, N. A., & Hakeem, A. Y. (2005). Intuition and autism: A possible role for von Economo neurons. *Trends in Cognitive Science, 9*(8), 367–373.

64. Hakeem, A. Y., Sherwood, C. C., Bonar, C. J., Butti, C., Hof, P. R., & Allman, J. M. (2009). Von Economo neurons in the elephant brain. *The Anatomical Record, 292*(2), 242–248.

65. Hof, P. R., & Van der Gucht, E. (2007). Structure of the cerebral cortex of the humpback whale, *Megaptera novaeangliae* (Cetacea, Mysticeti, Balaenopteridae). *The Anatomical Record, 290*(1), 1–31.

66. Butti, C., Sherwood, C. C., Hakeem, A. Y., Allman, J. M., & Hof, P. R. (2009).

Total number and volume of von Economo neurons in the cerebral cortex of cetaceans. *Journal of Comparative Neurology, 515*(2), 243–259.

67. Bystron, I., Rakic, P., Molnár, Z., & Blakemore, C. (2006). The first neurons of the human cerebral cortex. *Nature Neuroscience, 9,* 880–886.

CHAPTER 2:
THE PARALLEL AND DISTRIBUTED BRAIN

1. Galton, F. (1879). Psychometric experiments. *Brain, 2,* 149–162.

2. Caramazza, A., & Shelton, J. R. (1998). Domain-specific knowledge systems in the brain: The animate-inanimate distinction. *Journal of Cognitive Neuroscience, 10*(1), 1–34.

3. Boyer, P., & Barrett, H. C. (2005). Domain specificity and intuitive ontology. In D. M. Buss (Ed.), *The handbook of evolutionary psychology* (pp. 96–118). New York: Wiley.

4. Barrett, H. C. (2005). Adaptations to predators and prey. In D. M. Buss (Ed.), *The handbook of evolutionary psychology* (pp. 200–223). New York: Wiley.

5. Coss, R. G., Gusé, K. L., Poran, N. S., & Smith, D. G. (1993). Development of antisnake defenses in California ground squirrels (Spermophilus beecheyi): II. Microevolutionary effects of relaxed selection from rattlesnakes. *Behaviour, 124*(1–2), 137–164.

6. See: Stamm, J. S., &. Sperry, R. W. (1957). Function of corpus callosum in contralateral transfer of somesthetic discrimination in cats. *Journal of Comparative Physiological Psychology, 50*(2), 138–143; and Glickstein, M., & Sperry, R. W. (1960). Intermanual somesthetic transfer in split-brain rhesus monkeys. *Journal of Comparative Physiological Psychology, 53*(4), 322–327.

7. Akelaitis, A. J. (1945). Studies on the corpus callosum: IV. Diagnostic dyspraxia in epileptics following partial and complete section of the corpus callosum. *American Journal of Psychiatry, 101,* 594–599.

8. See: Gazzaniga, M. S., Bogen, J. E., & Sperry, R. W.(1962). Some functional effects of sectioning the cerebral commissures in man. *Proceedings of the National Academy of Sciences of the United States of America, 48*(10), 1765–1769; Gazzaniga, M. S., Bogen, J. E., & Sperry, R. W. (1963). Laterality effects in somesthesis following cerebral commissurotomy in man. *Neuropsychologia, 1,* 209–215; Gazzaniga, M. S., Bogen, J. E., & Sperry, R. W. (1965). Observations on visual perception after disconnection of the cerebral hemispheres in man. *Brain, 88,* 221–236; and Gazzaniga, M. S., Sperry, R. W. (1967). Language after section of the cerebral commissures. *Brain, 90,* 131–348.

9. Van Wagenen, W. P., & Herren, R. Y. (1940). Surgical division of commis-

sural pathways in the corpus callosum: Relation to spread of an epileptic attack. *Archives of Neurology and Psychiatry, 44*(4), 740–759.

10. Akelaitis, A. J. (1941). Studies on the corpus callosum: II. The higher visual
functions in each homonymous field following complete section of the corpus callosum. *Archives of Neurology and Psychiatry, 45*(5), 788–796.

11. Sperry, R. (1984). Consciousness, personal identity and the divided brain.
Neuropsychologia, 22(6), 661–673.

12. Kutas, M., Hillyard, S. A., Volpe, B. T., & Gazzaniga, M. S. (1990). Late
positive event-related potentials after commissural section in humans. *Journal of Cognitive Neuroscience, 2*(3), 258–271.

13. Gazzaniga, M. S., Bogen, J. E., & Sperry, R. W. (1967). Dyspraxia following
division of the cerebral commissures. *Archives of Neurology, 16*(6), 606–612.

14. See: Nass, R. D., & Gazzaniga, M. S.(1987). Cerebral lateralization and specialization in human central nervous system. In F. Plum (Ed.), *Handbook of
Physiology* (Sec. 1, Vol. 5, pp. 701–761). Bethesda, MD: American Physiological Society; and Zaidel, E. (1990). Language functions in the two hemispheres following cerebral commissurotomy and hemispherectomy. In F.
Boller & J. Grafman (Eds.), *Handbook of Neuropsychology* (Vol. 4, pp. 115–
150). Amsterdam: Elsevier.

15. Gazzaniga, M. S., & Smylie, C. S. (1990). Hemispheric mechanisms controlling voluntary and spontaneous facial expressions. *Journal of Cognitive Neuroscience, 2*(3), 239–245.

16. Sperry, R. W. (1968). Hemisphere deconnection and unity in conscious
awareness. *American Psychologist, 23*(10), 723–733.

17. Gazzaniga, M. S. (1972). One brain—two minds? *American Scientist, 60*(3),
311–317.

18. Sutherland, S. (1989). *The international dictionary of psychology.* New York:
Continuum.

19. MacKay, D. M. (1991). *Behind the eye.* Oxford: Basil Blackwell.

20. See: Phelps, E. A., & Gazzaniga, M. S. (1992). Hemispheric differences in
mnemonic processing: The effects of left hemisphere interpretation. *Neuropsychologia, 30*(3), 293–297; and Metcalfe, J., Funnell, M., & Gazzaniga,
M. S. (1995). Right-hemisphere memory superiority: Studies of a split-brain
patient. *Psychological Science, 6*(3), 157–164.

21. Nelson, M. E., & Bower, J. M. (1990). Brain maps and parallel computers.
Trends in Neurosciences, 13(10), 403–408.

22. Clarke, D. D., & Sokoloff, L. (1999). Circulation and energy metabolism of
the brain. In G. J. Siegel, B. W. Agranoff, R. W. Albers, S. K. Fisher, & M. D.
Uhler (Eds.), *Basic neurochemistry: Molecular, cellular and medical aspects*
(6th ed., pp. 637–670). Philadelphia: Lippincott-Raven.

23. Striedter, G. (2005). *Principles of brain evolution.* Sunderland, MA: Sinauer Associates, Inc.

24. Chen, B. L., Hall, D. H., & Chklovskii, D. B. (2006). Wiring optimization can relate neuronal structure and function. *Proceedings of the National Academy of Sciences of the United States of America, 103*(12), 4723–4728.

25. See: Hilgetag, C. C., Burns, G. A., O'Neill, M. A., Scannell, J. W., & Young, M. P. (2000). Anatomical connectivity defines the organization of clusters of cortical areas in the macaque monkey and the cat. *Philosophical Transactions of the Royal Society London B: Biological Sciences, 355*(1393), 91–110; Sporns, O., Tononi, G., & Edelman, G. M. (2002). Theoretical neuroanatomy and the connectivity of the cerebral cortex. *Behavioural Brain Research, 135*(1–2), 69–74; Sakata, S., Komatsu, Y., & Yamamori, T. (2005). Local design principles of mammalian cortical networks. *Neuroscience Research, 51*(3), 309–315.

26. Watts, D. J., & Strogatz, S. H. (1998). Collective dynamics of "small-world" networks. *Nature, 393*, 440–442.

27. See: Gazzaniga, M. S. (1989). Organization of the human brain. *Science, 245*(4921), 947–952; and Baynes, K., Eliassen, J. C., Lutsep, H. L., & Gazzaniga, M. S. (1998). Modular organization of cognitive systems masked by interhemispheric integration. *Science, 280*(5365), 902–905.

28. Volpe, B. T., Ledoux, J. E., & Gazzaniga, M. S. (1979). Information processing of visual stimuli in an "extinguished" field. *Nature, 282*(5740), 722–724.

29. Nicolis, G., & Rouvas-Nicolis, C. (2007). Complex systems. *Scholarpedia, 2*(11), 1473.

30. Amaral, L. A. N., & Ottino, J. M. (2004). Complex networks. Augmenting the framework for the study of complex systems. *European Physical Journal B, 38*(2), 147–162.

31. Varian, H. R. (2007). Position auctions. *International Journal of Industrial Organization, 25*(6), 1163–1178.

CHAPTER 3: THE INTERPRETER

1. Aglioti, S., DeSouza, J. F. X., & Goodale, M. A. (1995). Size-contrast illusions deceive the eye but not the hand. *Current Biology, 5*(6), 679–685.

2. Dehaene, S., Naccache, L., Le Clec'H, G., Koechlin, E., Mueller, M., Dehaene-Lambertz, G., . . . Le Bihan, D. (1998). Imaging unconscious semantic priming. *Nature, 395*, 597–600.

3. He, S., & MacLeod, D. I. A. (2001). Orientation-selective adaptation and tilt after-effect from invisible patterns. *Nature, 411*, 473–476.

4. Gazzaniga, M. S. (1989). Organization of the human brain. *Science, 245*(4921), 947–952.

5. Derks, P. L., & Paclisanu, M. I. (1967). Simple strategies in binary prediction by children and adults. *Journal of Experimental Psychology, 73*(2), 278–285.

6. Wolford, G., Miller, M. B., & Gazzaniga, M. S. (2000). The left hemisphere's role in hypothesis formation. *Journal of Neuroscience, 20*(6), RC64.

7. Kleck, R. E., & Strenta, A. (1980). Perceptions of the impact of negatively valued physical characteristics on social integration. *Journal of Personality and Social Psychology, 39*(5), 861–873.

8. Schachter, S., & Singer, J. E. (1962). Cognitive, social, and physiological determinants of emotional state. *Psychology Review, 69*, 379–399.

9. Miller, M. B., & Valsangkar-Smyth, M. (2005). Probability matching in the right hemisphere. *Brain and Cognition, 57*(2), 165–167.

10. Wolford, G., Miller, M. B., & Gazzaniga, M. S. (2004). Split decisions. In M. S. Gazzaniga (Ed.), *The Cognitive Neurosciences III* (pp. 1189–1199). Cambridge, MA: The MIT Press.

11. Corballis, P. (2003). Visuospatial processing and the right-hemisphere interpreter. *Brain and Cognition, 53*(2), 171–176.

12. Corballis, P. M., Fendrich, R., Shapley, R. M., & Gazzaniga, M. S. (1999). Illusory contour perception and amodal boundary completion: Evidence of a dissociation following callosotomy. *Journal of Cognitive Neuroscience, 11*(4), 459–466.

13. Corballis, P. M., Funnell, M. G., & Gazzaniga, M. S. (2002). Hemispheric asymmetries for simple visual judgments in the split brain. *Neuropsychologia, 40*(4), 401–410.

14. Corballis, M. C., & Sergent, J. (1988). Imagery in a commissurotomized patient. *Neuropsychologia, 26*(1), 13–26.

15. See: Funnell, M. G., Corballis, P. M., & Gazzaniga, M. S. (2003). Temporal discrimination in the split brain. *Brain and Cognition, 53*(2), 218–222; and Handy, T. C., Gazzaniga, M. S., & Ivry, R. B. (2003). Cortical and subcortical contributions to the representation of temporal information. *Neuropsychologia, 41*(11), 1461–1473.

16. Hikosaka, O., Miyauchi, S., & Shimojo, S. (1993). Focal visual attention produces illusory temporal order and motion sensation. *Vision Research, 33*(9), 1219–1240.

17. Tse, P., Cavanagh, P., & Nakayama, K. (1998). The role of parsing in high-level motion processing. In T. Watanabe (Ed.), *High-level motion processing: Computational, neurobiological, and psychophysical perspectives* (pp. 249–266). Cambridge, MA: The MIT Press.

18. Corballis, P. M., Funnell, M. G., & Gazzaniga, M. S. (2002). An investigation of the line motion effect in a callosotomy patient. *Brain and Cognition, 48*(2–3), 327–332.

19. Ramachandran, V. S. (1995). Anosognosia in parietal lobe syndrome. *Conciousness and Cognition, 4*(1), 22–51.

20. Hirstein, W., & Ramachandran, V. S. (1997). Capgras syndrome: A novel probe for understanding the neural representation of the identity and familiarity of persons. *Proceedings of the Royal Society B: Biological Sciences, 264*(1380), 437–444.

21. Doran, J. M. (1990). The Capgras syndrome: Neurological/neuropsychological perspectives. *Neuropsychology, 4*(1), 29–42.

22. Roser, M. E., Fugelsang, J. A., Dunbar, K. N., Corballis, P. M., & Gazzaniga, M. S. (2005). Dissociating processes supporting causal perception and causal inference in the brain. *Neuropsychology, 19*(5), 591–602.

23. Gazzaniga, M. S.(1983). Right hemisphere language following brain bisection: A 20-year perspective. *American Psychologist, 38*(5), 525–537.

24. Gazzaniga, M. S., & LeDoux, J. E. (1978). *The integrated mind*. New York: Plenum Press.

25. Roser, M., & Gazzaniga, M. S. (2004). Automatic brains—Interpretive minds. *Current Directions in Psychological Science, 13*(2), 56–59.

CHAPTER 4:
ABANDONING THE CONCEPT OF FREE WILL

1. Personal communication.

2. Fried, I., Katz, A., McCarthy, G., Sass, K. J., Williamson, P., Spencer, S. S., & Spenser, D. D. (1991). Functional organization of human supplementary motor cortex studied by electrical stimulation. *Journal of Neuroscience, 11*(11), 3656–3666.

3. Thaler, D., Chen, Y. C., Nixon, P. D., Stern, C. E., & Passingham, R. E. (1995). The functions of the medial premotor cortex. I. Simple learned movements. *Experimental Brain Research, 102*(3), 445–460.

4. Lau, H., Rogers, R. D., & Passingham, R. E. (2006). Dissociating response selection and conflict in the medial frontal surface. *NeuroImage, 29*(2), 446–451.

5. Lau, H. C., Rogers, R.D., & Passingham, R. E. (2007). Manipulating the experienced onset of intention after action execution. *Journal of Cognitive Neuroscience, 19*(1), 1–10.

6. Vohs, K. D., & Schooler, J. W. (2008). The value in believing in free will. Encouraging a belief in determinism increases cheating. *Psychological Science, 19*(1), 49–54.

7. See: Harmon-Jones, E., & Mills, J. (1999). *Cognitive dissonance: Progress on a pivotal theory in social psychology*. Washington, DC: American Psychologi-

cal Association; and Mueller, C. M., & Dweek, C. S. (1998). Intelligence praise can undermine motivation and performance. *Journal of Personality and Social Psychology, 75,* 33–52.

8. See: Baumeister, R. F., Bratslavsky, E., Muraven, M., & Tice, D. M. (1998). Ego depletion: Is the active self a limited resource? *Journal of Personality and Social Psychology, 4,* 1252–1265; Gailliot, M. T., Baumeister, R. F., DeWall, C. N., Maner, J. K., Plant, E. A., Tice, D. M., & Brewer, L. E. (2007). Self-control relies on glucose as a limited energy source: Willpower is more than a metaphor. *Journal of Personality and Social Psychology, 92,* 325–336; and Vohs, K. D., Baumeister, R. F., Schmeichel, B. J., Twenge, J. M., Nelson, N. M., & Tice, D. M. (2008). Making choices impairs subsequent self-control: A limited resource account of decision making, self-regulation, and active initiative. *Journal of Personality and Social Psychology, 94,* 883–898.

9. Baumeister, R. F., Masicampo, E. J., & DeWall, C. N. (2009). Prosocial benefits of feeling free: Disbelief in free will increases aggression and reduces helpfulness. *Personality and Social Psychology Bulletin, 35*(2), 260–268.

10. Dawkins, R. (2006). Edge.org, 1/1.

11. O'Connor, J. J., & Robertson, E. F. (2008). Edward Norton Lorenz. http://www.history.mcs.st-and.ac.uk/Biographies/Lorenz_Edward.html.

12. Feynman, R. (1998). *The meaning of it all.* New York: Perseus Books Group.

13. Bohr, M. (1937). Causality and complementarity. *Philosophy of Science, 4*(3), 289–298.

14. Quoted in: Isaacson, W. (2007). *Einstein: His Life and Universe.* New York: Simon & Schuster.

15. Pattee, H. H. (2001). Causation, control, and the evolution of complexity. In P. B. Andersen, P. V. Christiansen, C. Emmeche, & M. O. Finnerman (Eds.), *Downward causation: Minds, bodies and matter* (pp. 63–77). Copenhagen: Aarhus University Press.

16. Goldstein, J. (1999). Emergence as a construct: History and issues. *Emergence: Complexity and Organization, 1*(1), 49–72.

17. Laughlin, R. B. (2006). *A different universe: Reinventing physics from the bottom down.* New York: Basic Books.

18. Feynman, R. P., Leighton, R. B., & Sands, M.(1995). *Six easy pieces: Essentials of physics explained by its most brilliant teacher* (p. 135). New York: Basic Books.

19. Bunge, M. (2010). *Matter and mind: A philosophical inquiry* (p. 77). Dordrecht: Springer Verlag.

20. Libet, B., Wright, E. W., Feinstein, B., & Pearl, D. K. (1979). Subjective referral of the timing for a conscious sensory experience: A functional role for the somatosensory specific projection system in man. *Brain, 102*(1), 193–224.

21. Libet, B., Gleason, C. A., Wright, E. W., & Pearl, D. K. (1983). Time of conscious intention to act in relation to onset of cerebral activity (readiness-potential): The unconscious initiation of a freely voluntary act. *Brain, 106*(3), 623–642.

22. Soon, C. S., Brass, M., Heinze, H.-J. & Haynes, J.-D. (2008). Unconscious determinants of free decisions in the human brain. *Nature Neuroscience, 11*(5), 543–545.

23. Prinz, A. A., Bucher, D., & Marder, E. (2004). Similar network activity from disparate circuit parameters. *Nature Neuroscience, 7*(12), 1345–1352.

24. Anderson, P. W. (1972). More is different. *Science, 177*(4047), 393–396.

25. Locke, J. (1689). *An essay concerning human understanding* (1849 ed., p. 155). Philadelphia: Kay & Troutman.

26. Krakauer, D. Personal communication.

27. Bassett, D. S., & Gazzaniga, M. S. (2011). Understanding complexity in the human brain. *Trends in Cognitive Science*, in press.

CHAPTER 5: THE SOCIAL MIND

1. Legerstee, M. (1991). The role of person and object in eliciting early imitation. *Journal of Experimental Child Psychology, 51*(3), 423–433.

2. For a review, see: Puce, A., & Perrett, D. (2003). Electrophysiology and brain imaging of biological motion. *Philosophical Transactions of the Royal Society of London B: Biological Sciences, 358,* 435–446.

3. Heider, F., & Simmel, M. (1944). An experimental study of apparent behavior. *American Journal of Psychology, 57*(2), 243–259.

4. Premack, D., & Premack, A. (1997). Infants attribute value to the goal-directed actions of self-propelled objects. *Journal of Cognitive Neuroscience, 9*(6), 848–856.

5. Hamlin, J. K., Wynn, K., & Bloom, P. (2007). Social evaluation by preverbal infants. *Nature, 450,* 557–559.

6. Warneken, F., & Tomasello, M. (2007). Helping and cooperation at 14 months of age. *Infancy, 11*(3), 271–294.

7. Warneken, F., Hare, B., Melis, A. P., Hanus, D., & Tomasello, M. (2007). Spontaneous altruism by chimpanzees and young children. *PLoS Biology, 5*(7), 1414–1420.

8. Warneken, F., & Tomasello, M. (2006). Altruistic helping in human infants and young chimpanzees. *Science, 311*(5765), 1301–1303.

9. Liszkowski, U., Carpenter, M., Striano, T., & Tomasello, M. (2006). 12- and 18-month-olds point to provide information for others. *Journal of Cognition and Development, 7*(2), 173–187.

10. Warneken, F., & Tomasello, M. (2009). Varieties of altruism in children and chimpanzees. *Trends in Cognitive Science, 13*(9), 397–402.

11. Olson, K. R., & Spelke, E. S. (2008). Foundations of cooperation in young children. *Cognition, 108*(1), 222–231.

12. Melis, A. P., Hare, B., & Tomasello, M. (2008). Do chimpanzees reciprocate received favours? *Animal Behaviour, 76*(3), 951–962.

13. Rakoczy, H., Warneken, F., & Tomasello, M. (2008). The sources of normativity: Young children's awareness of the normative structure of games. *Developmental Psychology, 44*(3), 875–881.

14. Stephens, G. J., Silbert, L. J., & Hasson, U. (2010). Speaker-listener neural coupling underlies successful communication. *Proceedings of the National Academy of Sciences of the United States of America, 107*(32), 14425–14430.

15. Jolly, A. (1966). Lemur and social behavior and primate intelligence. *Science, 153*(3735), 501–506.

16. Byrne, R. W., & Whiten, A. (1988). *Machiavellian intelligence.* Oxford: Clarendon Press.

17. Byrne, R. W., & Corp, N. (2004). Neocortex size predicts deception rate in primates. *Proceedings of the Royal Society B: Biological Sciences, 271*(1549), 1693–1699.

18. Moll, H., & Tomasello, M. (2007). Cooperation and human cognition: The Vygotskian intelligence hypothesis. *Philosophical Transactions of the Royal Society B: Biological Sciences, 362*(1480), 639–648.

19. Dunbar, R. I. M. (1998). The social brain hypothesis. *Evolutionary Anthropology, 6*(5), 178–190.

20. Dunbar, R. I. M. (1993). Coevolution of neocortical size, group size and language in humans. *Behavioral and Brain Sciences, 16*(4), 681–735.

21. Hill, R. A., & Dunbar, R. I. M. (2003). Social network size in humans. *Human Nature, 14*(1), 53–72.

22. Roberts, S. G. B., Dunbar, R. I. M., Pollet, T. V., & Kuppens, T. (2009). Exploring variation in active network size: Constraints and ego characteristics. *Social Networks, 1*(2), 138–146.

23. Dunbar, R. I. M. (1996). *Grooming, gossip, and the evolution of language.* Cambridge, MA: Harvard University Press.

24. Papineau, D. (2005). Social learning and the Baldwin effect. In A. Zilhão (Ed.), *Evolution, rationality and cognition: A cognitive science for the twenty-first century* (pp. 40–60). New York: Routledge.

25. Baldwin, J. M. (1896). A new factor in evolution. *The American Naturalist, 30*(354), 441–451.

26. Krubitzer, L., & Kaas, J. (2005). The evolution of the neocortex in mammals:

How is phenotypic diversity generated? *Current Opinion in Neurobiology, 15*(4), 444–453.

27. Lewontin, R. C. (1982). Organism and environment. In H. C. Plotkin (Ed.), *Learning, development and culture: Essays in evolutionary epistemology* (pp. 151–171). New York: Wiley.

28. Odling-Smee, F. J., Laland, K. N., & Feldman, M. W. (2003). Niche construction: The neglected process in evolution. Retrieved from http://www.nicheconstruction.com/.

29. Flack, J. C., de Waal, F. B. M., & Krakauer, D. C. (2005). Social structure, robustness, and policing cost in a cognitively sophisticated species. *The American Naturalist, 165*(5), E126–E139.

30. Flack, J. C., Krakauer, D. C., & de Waal, F. B. M. (2005). Robustness mechanisms in primate societies: A perturbation study. *Proceedings of the Royal Society B: Biological Sciences, 272*(1568), 1091–1099.

31. Belyaev, D. (1979). Destabilizing selection as a factor in domestication. *Journal of Heredity, 70*(5), 301–308.

32. Hare, B., Plyusnina, I., Ignacio, N., Schepina, O., Stepika, A., Wrangham, R., & Trut, L. (2005). Social cognitive evolution in captive foxes is a correlated by-product of experimental domestication. *Current Biology, 15*(3), 226–230.

33. Allport, F. H. (1924). *Social psychology.* Boston: Houghton Mifflin.

34. Emler, N. (1994). Gossip, reputation, and adaptation. In R. F. Goodman & A. Ben-Ze'ev (Eds.), *Good gossip* (pp. 117–138). Lawrence, KS: University Press of Kansas.

35. Call, J., & Tomasello, M. (2008). Does the chimpanzee have a theory of mind? 30 years later. *Trends in Cognitive Science, 12*(5), 187–192.

36. Bloom. P., & German, T. P. (2000). Two reasons to abandon the false belief task as a test of theory of mind. *Cognition, 77*(1), B25–B31.

37. Buttelmann, D., Carpenter, M., & Tomasello, M. (2009). Eighteen-month-old infants show false belief understanding in an active helping paradigm. *Cognition, 112*(2), 337–342.

38. See: Baron-Cohen, S. (1995). *Mindblindness: An essay on autism and theory of mind.* Cambridge, MA: The MIT Press; and Baron-Cohen, S., Leslie, A. M., & Frith, U. (1985). Does the autistic child have a "theory of mind"? *Cognition, 21*(1), 37–46.

39. Rizzolatti, G., Fadiga, L., Gallese, V., & Fogassi, L. (1996). Premotor cortex and the recognition of motor actions. *Cognitive Brain Research, 3*(2), 131–141.

40. Fadiga, L., Fogassi, L., Pavesi, G., & Rizzolatti, G. (1995). Motor facilitation during action observation: A magnetic stimulation study. *Journal of Neurophysiology, 73*(6), 2608–2611.

41. Singer, T., Seymour, B., O'Doherty, J., Kaube, H., Dolan, R. J., & Frith, C. D. (2004). Empathy for pain involves the affective but not sensory components of pain. *Science, 303*(5661), 1157–1162.

42. Jackson, P. L., Meltzoff, A. N., & Decety, J. (2005). How do we perceive the pain of others? A window into the neural processes involved in empathy. *NeuroImage, 24*(3), 771–779.

43. Dimberg, U., Thunberg, M., & Elmehed, K. (2000). Unconscious facial reactions to emotional facial expressions. *Psychological Science, 11*(1), 86–89.

44. Chartrand, T. L, & Bargh, J. A. (1999). The chameleon effect: The perception-behavior link and social interaction. *Journal of Personality and Social Psychology, 76*(6), 893–910.

45. Giles, H., & Powesland, P. F. (1975). *Speech style and social evaluation.* London: Academic Press.

46. For a review see: Chartrand, T. L., Maddux, W. W., & Lakin, J. L. (2005). Beyond the perception-behavior link: The ubiquitous utility and motivational moderators of nonconscious mimicry. In R. R. Hassin, J. S. Uleman, & J. A. Bargh (Eds.), *The new unconscious* (pp. 334–361). New York: Oxford University Press.

47. van Baaren, R. B., Holland, R. W., Kawakami, K., & van Knippenberg, A. (2004). Mimicry and prosocial behavior. *Psychological Science, 15*(1), 71–74.

48. Chaiken, S. (1980). Heuristic versus systematic information processing and the use of source versus message cues in persuasion. *Journal of Personality and Social Psychology, 39*(5), 752–766.

49. Hatfield, E., Cacioppo, J. T., & Rapson, R. L. (1993). Emotional contagion. *Current Directions in Psychological Sciences, 2*(3), 96–99.

50. Lanzetta, J. T., & Englis, B. G. (1989). Expectations of cooperation and competition and their effects on observers' vicarious emotional responses. *Journal of Personality and Social Psychology, 56*(4), 543–554.

51. Bourgeois, P., & Hess, U. (1999). Emotional reactions to political leaders' facial displays: A replication. *Psychophysiology, 36*, S36.

52. Bourgeois, P., & Hess, U. (2007). The impact of social context on mimicry. *Biological Psychology, 77*(3), 343–352.

53. Yabar, Y., Cheung, N., Hess, U., Rochon, G., & Bonneville-Hébert, M. (2001). *Dis-moi si vous êtes intimes, je te dirais si tu mimes* [Tell me if you're intimate and I'll tell you if you'll mimic]. Paper presented at the 24th Annual Meeting of the Société Québécoise pour la Recherche en Psychologie, October 26–28. Chicoutimi, Canada.

54. de Waal, F. (2001). *The ape and the sushi master: Cultural reflections of a primatologist.* New York: Basic Books.

55. See: Baner, G., & Harley, H. (2001). The mimetic dolphin [Peer commen-

tary on the paper, "Culture in whales and dolphins" by L. Rendall & H. Whitehead]. *Behaviorial and Brain Sciences, 24,* 326–327.

56. Visalberghi, E., & Fragaszy, D. M. (1990). Do monkeys ape? In S. T. Parker & K. R. Gibson (Eds.), *Language and intelligence in monkeys and apes* (pp. 247–273). Cambridge: Cambridge University Press; and Whiten, A., & Ham, R. (1992). On the nature and evolution of imitation in the animal kingdom: Reappraisal of a century of research. In P. J. B. Slater, J. S. Rosenblatt, C. Beer, & M. Milinski (Eds.), *Advances in the study of behavior* (pp. 239–283). New York: Academic Press.

57. Kumashiro, M., Ishibashi, H., Uchiyama, Y., Itakura, S., Murata, A., & Iriki, A. (2003). Natural imitation induced by joint attention in Japanese monkeys. *International Journal of Psychophysiology, 50*(1–2), 81–99.

58. Hume, D. (1777). *An enquiry concerning the principles of morals* (1960 ed., p. 2). La Salle, IL: Open Court.

59. Brown, D. E. (1991). *Human universals.* New York: McGraw-Hill.

60. Haidt, J. (2010). Morality. In S. T. Fiske, D. T. Gilbert, & G. Lindzey (Eds.), *Handbook of social psychology* (5th ed., Vol. 2, pp. 797–832). Hoboken, NJ: Wiley.

61. Haidt, J. (2001). The emotional dog and its rational tail: A social intuitionist approach to moral judgment. *Psychological Review, 108*(4), 814–834.

62. Haidt, J., & Bjorklund, F. (2008). Social intuitionists answer six questions about moral psychology. In W. Sinnott-Armstrong (Ed.), *Moral psychology* (Vol. 2, pp. 181–217). Cambridge, MA: The MIT Press.

63. Westermarck, E. A. (1891). *The History of Human Marriage.* New York: Macmillan.

64. Shepher, J. (1983). *Incest: A biosocial view.* Orlando, FL: Academic Press; and Wolf, A. P. (1970). Childhood association and sexual attraction: A further test of the Westermarck hypothesis. *American Anthropologist, 72*(3), 864–874.

65. Lieberman, D., Tooby, J., & Cosmides, L. (2002). Does morality have a biological basis? An empirical test of the factors governing moral sentiments relating to incest. *Proceedings of the Royal Society B: Biological Sciences, 270*(1517), 819–826.

66. Greene, J. D., Sommerville, R. B., Nystrom, L. E., Darley, J. M., & Cohen, J. D. (2001). An fMRI investigation of emotional engagement in moral judgment. *Science, 293*(5537), 2105–2108.

67. Hauser, M. (2006). *Moral minds.* New York: HarperCollins.

68. Koenigs, M., Young, L., Adolphs, R., Tranel, D., Cushman, F., Hauser, M., & Damasio, A. (2007). Damage to the prefrontal cortex increases utilitarian moral judgements. *Nature, 446,* 908–911.

69. Pinker, S. (2008, January 13). The moral instinct. *The New York Times*. Retrieved from http:www.nytimes.com.

70. Haidt, J., & Joseph, C. (2004). Intuitive ethics: How innately prepared intuitions generate culturally variable virtues. *Dædalus, 133*(4), 55–66; and Haidt, J., & Bjorklund, F. (2008). Social intuitionists answer six questions about moral psychology. In W. Sinnott-Armstrong (Ed.), *Moral psychology* (Vol. 2, pp. 181–217). Cambridge, MA: The MIT Press.

71. Darwin, C. (1871). The descent of man. In M. Adler (Ed.), *Great books of the western world* (1952 ed., Vol. 49, p. 322). Chicago: Encyclopædia Britannica.

72. Knoch, D., Pascual-Leone, A., Meyer, K., Treyer, V., & Fehr, E. (2006). Diminishing reciprocal fairness by disrupting the right prefrontal cortex. *Science, 314*(5800), 829–832.

73. Anderson, S. W., Bechara, A., Damasio, H., Tranel, D., & Damasio, A. R. (1999). Impairment of social and moral behavior related to early damage in human prefrontal cortex. *Nature Neuroscience, 2*(11), 1032–1037.

CHAPTER 6: WE ARE THE LAW

1. Van Biema, D., Drummond, T., Faltermayer, C., & Harrison, L. (1997, March 3). A recurring nightmare. *Time*. Retrieved from http://www.time.com.

2. Spake, A. (1997, March 5). Newsreal: The return of Larry Singleton. *Salon*. Retrieved from http://www.salon.com.

3. Puit, G. (2002, January 6). 1978 Mutilation: Family relieved by Singleton's death. *Review Journal*. Retrieved from http://crimeshots.com/VincentNightmare.html.

4. Taylor, M. (2002, January 1). Lawrence Singleton, despised rapist, dies / He chopped off teenager's arms in 1978. *San Francisco Chronicle*. Retrieved from http:www.sfgate.com.

5. Harrower, J. (1998). *Applying psychology to crime*. Hillsdale, NJ: Lawrence Erlbaum Associates.

6. Hackett, R. (2003, January 30). A victim, a survivor, an artist. *Seattle Post-Intelligencer*. Retrieved from http://www.seattlepi.com/local/106424_maryvincent30.shtml.

7. Nisbett, R. E., Peng, K., Choi, I., & Norenzayan, A. (2001). Culture and systems of thought: Holistic versus analytic cognition. *Psychological Review, 108*(2), 291–310.

8. Nisbett, R. E. (2003). *The geography of thought: How Asians and Westerners think differently and why* (pp. 2–3, 5). New York: Free Press.

9. Hedden, T., Ketay, S., Aron, A., Markus, H. R., & Gabrieli, J. (2008).

Cultural influences on neural substrates of attentional control. *Psychological Science 19*(1), 12–17.

10. Uskul, A. K., Kitayama, S., & Nisbett, R. E. (2008). Ecocultural basis of cognition: Farmers and fishermen are more holistic than herders. *Proceedings of the National Academy of Sciences of the United States of America, 105*(25), 8552–8556.

11. Kim, H. S., Sherman, D. K, Taylor, S. E., Sasaki, J. Y., Chy, T. Q., Ryu, C., Suh, E. M., & Xu, J. (2010). Culture, serotonin receptor polymorphism and locus of attention. *Social Cognitive & Affective Neuroscience, 5,* 212–218.

12. Personal communication.

13. United Kingdom House of Lords decisions. Daniel M'Naghten's case. May 26, June 19, 1843. Retreived from http://www.bailii.org/uk/cases/UKHL/1843/J16.html.

14. Weisberg, D. S., Keil, F. C., Goodstein, J., Rawson, E., & Gray, J. R. (2008). The seductive allure of neuroscience explanations. *Journal of Cognitive Neuroscience, 20,* 470–477.

15. Shariff, A. F., Greene, J. D., Schooler, J. W. (submitted). His brain made him do it: Encouraging a mechanistic worldview reduces punishment.

16. Staff working paper (2004). An overview of the impact of neuroscience evidence in criminal law. *The President's council on Bioethics.* Retrieved from http://bioethics.georgetown.edu/pcbe/background/neuroscience_evidence.html.

17. Scalia, A. (2002). *Akins v. Virginia* (00-8452) 536 U.S. 304. Retrieved August 9, 2010, from http://www.law.cornell.edu/supct/html/00-8452.Z.

18. Snead. O. C. (2006). Neuroimaging and the courts: Standards and illustrative case index. *Report for Emerging Issues in Neuroscience Conference for State and Federal Judges.* Retrieved from http://webcache.googleusercontent.com/search?q=cache:CTy_7pLokKYJ:www.ncsconline.org/d_research/stl/dec06/Snead%2520Presentation%2520%28AAAS%2520-%2520modified%29.doc+simon+pirela&cd=10&hl=en&ct=clnk&gl=us&client=firefox-a.

19. Talairach, P. T., & Tournoux, P. (1988). *Co-planar stereotaxic atlas for the human brain: 3-D proportional system: An approach to cerebral imaging* (p. vii). New York: Thieme Medical Publishers.

20. Miller, M. B., van Horn, J. D., Wolford, G. L., Handy, T. C., Valsangkar-Smyth, M., Inati, S., . . . Gazzaniga, M. S. (2002). Extensive individual differences in brain activations associated with episodic retrieval are reliable over time. *Journal of Cognitive Neuroscience, 14*(8), 1200–1214.

21. Doron, C., & Gazzaniga, M. S. (2009). Neuroimaging techniques offer new perspectives on callosal transfer and interhemispheric communication. *Cortex, 44*(8), 1023–1029.

22. Putman, M. C., Steven, M. S., Doron, C., Riggall, A. C., & Gazzaniga, M. S. (2009). Cortical projection topography of the human splenium: Hemispheric asymmetry and individual difference. *Journal of Cognitive Neuroscience*, 22(8), 1662–1669.

23. Desmurget, M., Reilly, K. T., Richard, M., Szathmari, A., Mottolese, C., & Sirigu, A. (2009). Movement intention after parietal cortex stimulation in humans. *Science*, 324(811), 811–813.

24. Brass, M., & Haggard, P. (2008). The what, when, whether model of intentional action. *Neuroscientist*, 14(4), 319–325.

25. Brass, M., & Haggard, P. (2007). To do or not to do: The neural signature of self-control. *Journal of Neuroscience*, 27(34), 9141–9145.

26. Kuhn, S., Haggard, P., & Brass, M. (2009). Intentional inhibition: How the "veto-area" exerts control. *Human Brain Mapping*, 30(9), 2834–2843.

27. Schauer, F. (2010). Neuroscience, lie-detection, and the law: Contrary to the prevailing view, the suitability of brain-based lie-detection for courtroom or forensic use should be determined according to legal and not scientific standards. *Trends in Cognitive Science*, 14(3), 101–103.

28. Bond, C. F., & De Paulo, B. M. (2006). Accuracy of deception judgments. *Personality and Social Psychology Review*, 10, 214–234.

29. Meisser, C. A., & Bigham, J. C. (2001). Thirty years of investigating the own-race bias in memory for faces: A meta-analytic review. *Psychology, Public Policy, and Law*, 7(1), 3–35.

30. Connors, E., Lundregar, T., Miller, N., & McEwan, T. (1996). *Convicted by juries, exonerated by science: Case studies in the use of DNA evidence to establish innocence after trial.* Washington, DC: National Institute of Justice.

31. Turk, D. J., Handy, T. C., & Gazzaniga, M. S. (2005). Can perceptual expertise account for the own-race bias in face recognition? A split-brain study. *Cognitive Neuropsychology*, 22(7), 877–883.

32. Harris, L. T., & Fiske, S. T. (2006). Dehumanizing the lowest of the low: Neuroimaging responses to extreme out-groups. *Psychological Science*, 17(10), 847–853.

33. Wilkinson, R. A. (1997). A shifting paradigm: Modern restorative justice principles have their roots in ancient cultures. *Corrections Today*. Dec. Retrieved from http://www.drc.state.oh.us/web/Articles/article28.htm.

34. Sloane, S., & Baillargeon, R. (2010). *2.5-Year-olds divide resources equally between two identical non-human agents.* Poster session presented at the annual meeting of the International Society of Infant Studies, Baltimore, MD.

35. Geraci, A., & Surian, L. (2010). *Sixteen-month-olds prefer agents that perform equal distributions.* Poster session presented at the annual meeting of the International Society of Infant Studies, Baltimore, MD.

36. He, Z., & Baillargeon, R. (2010). *Reciprocity within but not across groups: 2.5-year-olds' expectations about ingroup and outgroup agents.* Poster session presented at the annual meeting of the International Society of Infant Studies, Baltimore, MD.

37. Vaish, A., Carpenter, M., & Tomasello, M. (2010). *Moral mediators of young children's prosocial behavior toward victims and perpetrators.* Poster session presented at the annual meeting of the International Society of Infant Studies, Baltimore, MD.

38. Harris, P. L., & Nunez, M. (1996). Understanding permission rules by preschool children. *Child Development, 67*(4), 1572–1591.

39. Hamlin, J., Wynn, K., Bloom, P., & Mahagan, N., Third-party reward and punishment in young toddlers. (Under review)

40. Carlsmith, K. M. (2006). The roles of retribution and utility in determining punishment. *Journal of Experimental Social Psychology, 42,* 437–451.

41. Darley, J. M., Carlsmith, K. M., Robinson, P. H. (2000). Incapacitation and just deserts as motives for punishment. *Law and Human Behavior, 24,* 659–683.

42. Carlsmith, K. M., & Darley, J. M. (2008). Psychological aspects of retributive justice. In M. P. Zanna (Ed.), *Advances in experimental social psychology* (Vol. 40, pp. 193–236). San Diego, CA: Elsevier.

43. Carlsmith, K. M. (2008). On justifying punishment: The discrepancy between works and actions. *Social Justice Research, 21,* 119–137.

44. Buckholtz, J. W., Asplund, C. L., Dux, P. E., Zald, D. H., Gore, J. C., Jones, O. D., & Marois, R. (2008). The neural correlates of third-party punishment. *Neuron, 60,* 930–940.

45. Richards, J. R. (2000). *Human nature after Darwin* (p. 210). New York: Routledge.

46. Boyd, R., Gintis, H., Bowles, S., & Richerson, P. J. (2003). The evolution of altruistic punishment. *Proceedings of the National Academy of Sciences of the United States of America, 100*(6), 3531–3535.

INDEX